SHELL
SHELL
SHELL
SHELL
EMERGENCY
AAA
SERVICE
CONTAINS
LEAD
OLIVER
88
88

OLIVER
66

CLASSIC OLIVER TRACTORS

HISTORY, MODELS, VARIATIONS, & SPECIFICATIONS 1897–1976

SHERRY SCHAEFER

SECOND EDITION

OCTANE PRESS

Octane Press, Second Edition
First printing, March 2019
Second printing, May 2022
Voyageur Press, First Edition, July 2009

ISBN: 978-1-937747-99-2
LCCN: 2018959792

Cover and Interior Design by Tom Heffron
Copyedited by Karen O'Brien
Proofread by Jennifer Grecsek

On the cover: Under new ownership of the White Motor Company in 1960, Oliver introduced the 100 series tractors. The largest of those, the GM-powered 1900, made the company a contender in the high-horsepower line with the rest of the industry. *Super T*

On the frontispiece: Oliver was ahead of its time when it introduced the streamlined styling in 1937. This automotive design offered a sleek and modern look that protected the engine and its components from the elements. Although the streamlined styling was replaced in 1948 by the Fleetline styling, the early 88 wore it well during its early production in 1947 and 1948. *Super T*

On the title page: Known as the "three beauties", the 66, 77, and 88 were the foundation of the Fleetline series. Providing a model for the small, medium, or larger acreage farmer, the models were designed to share various components. The interchangeability of their parts allowed the dealer to stock fewer but still have enough on hand to keep the equipment rolling. *Super T*

On the backcover: As iconic as the Mail Pouch tobacco barn, the 60 was a staple on the farm of the Oliver user. The faithful "hired hand" was turned to for garden work, small fieldwork, raking hay, or pulling the feed wagon. An economical and reliable workhorse, this was the Shetland pony of the farm and many youngsters learned to drive on a 60. *Super T*

OCTANE PRESS

octanepress.com

Octane Press is based in Austin, Texas

Printed in the US

CONTENTS

THE OLIVER R 40
MAN'FD BY
OLIVER CHILLED PLOW WORKS.
SOUTH BEND. IND. U.S.A.
THE OLIVER

CHAPTER 1

THE PLANTS OF OLIVER

The Oliver Corporation's roots run deep in the agricultural industry. This once-small company started with a young man and an idea that would lighten the load of the farmer. Through the years, the company grew and acquired other companies with that same philosophy. The result was nearly a dozen companies operating under the Oliver flag to become a strong force in the farm market.

The original Oliver business started in 1855 when James Oliver, a young Scottish immigrant, bought one quarter of the interest in the South Bend Foundry located in Mishawaka, Indiana, on the St. Joseph River. The major shareholder was Emsley Lamb. Just one year later, Lamb sold his half to Oliver and Harvey Little. The company was renamed the South Bend Iron Company but was known as Oliver and Little to some in the area.

At this point, a majority of the foundry's work was casting iron columns, windowsills, and other castings for building purposes, but James Oliver made it his ambition to manufacture the best plow in the world.

In 1857, James Oliver and Harvey Little applied for their first patent that dealt with the chilling of plowshares. This new method of casting gave the share a hardened surface that was durable and long lasting. Oliver was so confident with his invention that he issued a challenge through the newspaper, "500 dollar reward to any furnace man in the United States that could chill or harden a plowshare with equal success without infringing on the method developed by Oliver & Little." This was a true testimony of their faith in this method. It was 11 years before this method of chilling was applied to the moldboard. During this lapse, Oliver was constantly experimenting to further perfect the process.

Although 1857 was tough for the Oliver & Little enterprise, three men were hired. The positions were of different specialties, and the pay ranged from $1.00

The Oliver 40 walking plow was sold around the world for nearly a century. Color photographs of it were rarely used, but this rare catalog shot depicts the actual color scheme used in the early 1900s. *Author's collection*

to $1.75 per day. Money rarely changed hands in the matter of wages; instead, workers received most of their pay in exchange for goods at various retailers in town. Local merchants were usually indebted to the foundries and were more than willing to exchange commodities such as flour, lumber, and shoes for services. Oliver & Little used the trade to pay its workers.

The Panic of 1857 sent the country into economic depression. People were pulling their money out of banks. Grain prices were dropping. Manufacturing plants were full of inventory, and no one was buying machinery. If a farmer needed a new plow, he was going to make the old one last another year. Plows sold from $6.25 to $9.00; yet the company sold less than fifty of them that year. It was not a prosperous year, but it was not disastrous. During the lull, many improvements were made to the factory, and word started to spread about its successful chilled plow.

James Oliver was fond of attending fairs. He began to travel to various fairs to show the company's different models of plows, and his efforts were rewarded. The company sold more than two hundred plows in 1858, which was quite an improvement over the previous year since the country was still suffering through the 1857 depression (that lasted three years).

Oliver's primary competition came from the St. Joseph Iron Works, which was running neck-in-neck with its plow business. Both companies had won awards for their different types of plows. Oliver was still convinced it had the better overall plow, so it put forth a challenge to the St. Joseph Iron Works in the form of a plowing match. The challenge was advertised in the local paper, and all who were interested were invited to attend.

The match was to test sod plow versus sod plow and stubble plow versus stubble plow. On the designated day, both companies and their representatives showed up, but at the last minute, one of the St. Joseph representatives objected to one of the judges. The judge in question was removed and replaced to satisfy all. Just as the contest was about to begin, the company refused to test its sod plow by saying the ground was not suitable.

Though it would appear that the match was off, Oliver decided to go ahead and test its plow along with a St. Joseph Hoosier plow that Oliver had purchased. Oliver's Indiana Sod Plow had 400 to 425 pounds of draft, while the Hoosier plow had 500 pounds. Oliver then tested its own Indiana Stubble Plow, which registered a draft of 525, against the Hoosier Plow, which showed 600 pounds of draft. Oliver showed that it was superior.

St. Joseph Iron Works objected to this display, and a war of words took place in the newspapers. Oliver boasted of its superiority, while St. Joseph Iron Works claimed that the judges were biased. These barbs went back and forth for a while, but no damage was ever done to the Oliver business, as the plows were proof enough.

In 1860, Oliver turned out one of its most successful plows ever—the Model 40. With a price tag of $6.50, it was the plow model that put the company out front. By 1911, more than one million Model 40s had been sold, and its production continued well into the 1940s.

The company had six employees in 1860, not counting the owners. By 1865, forty men were on the payroll. Wages had increased, and the average man was paid $2.50 a day. One of those employees was Joseph Doty Oliver, the fourteen-year-old son of James Oliver. He began his work in the foundry by threading nuts and learning the plow trade from the ground up. Joseph, known as J.D., left the foundry for a few years to attend college. When he returned, he became the company bookkeeper and concentrated his efforts on running the office. J.D. became the organizer and financial brains of the company, while his father remained the inventor and builder. It was a perfect combination that sent the company straight to the top of the plow business.

Oliver had changed partners once again, and the company was now known as Oliver, Bissell, and Company. The company was growing, and more buildings had to be erected to house supplies and inventory. By 1872, Oliver was managing the largest foundry in northern Indiana. The company was building forty plows per day in addition to its contracted casting work for Studebaker Wagons and Singer Sewing Machine. In 1874, it built 14,976 plows. The small foundry that James Oliver had acquired in 1855 was no longer able to house the growing company and meet the sales demand.

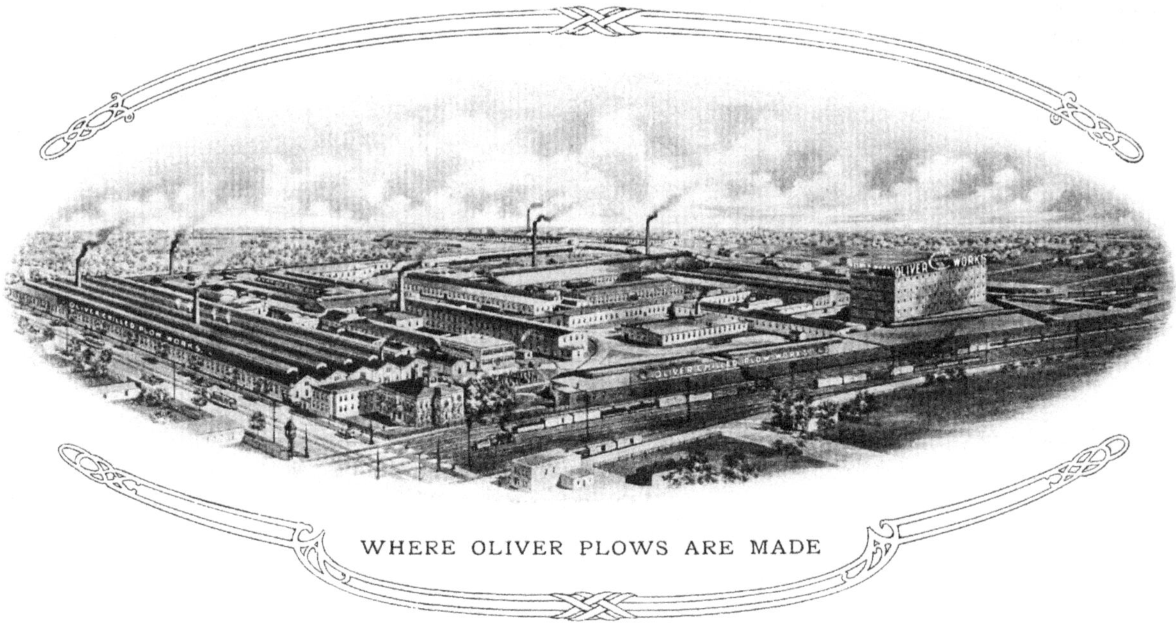

This aerial photo depicts the Oliver Chilled Plow Works in South Bend, Indiana, around 1910. At this time, the plant was powered by a hydro-electric power plant that generated 3,600 horsepower. *Author's collection*

Oliver's plows had proven themselves; if the company was going to continue in its success, relocation was necessary. In 1875, James purchased 32 acres on which he planned to build a new factory. The next year was hectic with construction, relocation, and trying to keep the plow business operating during the building process. The new plant was the largest plant in the world dedicated specifically to the manufacture of plows, with 450 employees and a capacity of 300 plows per day. Five buildings were built in the first year of construction. More than four million bricks were used along with 1,320 feet of main shafting, thirty-one woodworking machines, and seventy-five grindstones. The total floor space was 200,000 square feet.

A 600-horsepower Harris steam engine powered the plant, and the pump house had a capacity of 350,000 gallons of water every ten hours. In addition to the new plant, Oliver had branch houses located in Indianapolis, Indiana; Mansfield, Ohio; and Dallas, Texas.

The South Bend Iron Company continued to flourish and grow into the next century. The orders for Oliver plows were coming in faster than they could be turned out. In 1901, the company was reorganized and incorporated to take over the business of the iron works. Thus, the Oliver Chilled Plow Works was born with a capital stock value of $500,000. At that point, the company was 100 percent owned by the Oliver family.

On a single day in 1902, the foundry turned out a record of nine hundred plows and nine thousand points. Sales were no longer limited to the United States. Oliver had several export salesmen who traveled extensively to market the Oliver plow all over the world. Records show that in 1908 the output of the plant was approximately three hundred thousand plows a year with roughly 220 employees. With the prosperous sales and production levels, the company took its biggest hit in 1908 when James Oliver passed away at the age of 84.

By 1909, J.D. Oliver was running the company founded by his late father. Sales were at an all-time high, and J.D. put a plan in motion both to double the size of the South Bend factory and to build a new plant in Hamilton, Ontario. The first shipment of plows left the Hamilton plant in 1911. The International Harvester Company handled Oliver plow distribution in Canada. This continued until 1919 when Oliver sold the plant to IH.

The ground occupied by the Oliver plant in South Bend was in excess of 58 total acres with 35 acres of floor space by 1910. Five miles of railroad tracks were within the plant, and the private water works had a daily capacity of 1.5 million gallons. The plant was now capable of manufacturing five hundred thousand plows a year, and these plows were leaving their mark on the landscape.

Desolate prairie land was turned into fertile farm land thanks to the innovations of James Oliver. Horse farming was slowly being transformed to power farming, but horse plows and tractor plows were two different pieces of equipment. In 1919, anyone with a little imagination was trying to manufacture a tractor, and it became known as "The Year of the Tractor." J.D. Oliver teamed up with Henry Ford to supply plows to match up to Ford's equipment. With this union, Ford could offer a complete package to the farmer—a proven tractor and a proven plow to go along with it.

The current factory was massive in size and capacity, though there still wasn't enough room to start building tractor-drawn equipment because of all the products that Oliver offered. The only natural action in Oliver's eyes was to build yet another plant. This plant was known as South Bend Plant No. 2. Located just a few blocks away from Plant No. 1, this new plant was dedicated exclusively to the manufacture of tractor-drawn equipment.

Oliver took on its most extensive expansion program during the 1920s. Not only had it built a plant on 75 acres in Canada and a second plant in South Bend, but the company had fourteen branch houses in thirteen different states from New York to California to Texas.

J.D. was getting older and enlisted his son-in-law, Charles Frederick Cunningham, to design and oversee the building of the second plant. Charles took as much pride in this new project as James had in perfecting the plow. Cunningham's son, Ollie, recalls that his father would ask him if he wanted to go see daddy's baby. As an excited youngster, Ollie would jump in the car and ride to the new plant with his father. Upon their arrival, Ollie would look all over for a little baby but could never find one.

Plant No. 2 was built to cater to the needs of Henry Ford and his tractors. Upon completion of the plant, however, Henry announced that he was going to devote his time to the automobile and was getting out of the tractor business. This move left Oliver's new plant without much of a purpose. J.D. was livid, and according to his grandson, Ollie, he never bought another Ford vehicle as long as he lived.

Hart-Parr: Charles City, Iowa

While James Oliver was basking in the glow of his successful plow, two young college boys were busy designing a gasoline combustion engine in Madison, Wisconsin. Charles William Hart and Charles Henry Parr had both enrolled in the mechanical engineering program at the University of Wisconsin. The two first met in 1893 at the school when they struck up a conversation and realized they had the same interest. They collaborated on a college thesis dealing with the internal combustion engine.

While still in school, the two operated a small shop where they repaired farm equipment and experimented with building engines. After graduating with honors, they incorporated as the Hart-Parr Company beginning with $24,000 in capital. They purchased a small lot for $3,000 and erected a 31 x 56-foot two-story building. Hart and Parr began building their successful line of Hart-Parr engines in this building.

In 1900, Hart and Parr decided to build a foundry so that they could build their own castings. It soon became evident that they needed more room to handle the volume of business that was coming their way, but Madison was not holding open the gates. Land was expensive there, and the city fathers were not in favor of a manufacturing plant. They preferred to keep their town more residential.

While Hart was visiting his parents in Charles City during 1901, he spoke of the dilemma he and Parr were facing. Hart's father approached the Ellis brothers, who were leading businessmen in Charles City, about helping his son relocate the Hart-Parr business. The city was all for industrial growth and put together an incentive package to entice the company to relocate to Charles City.

On June 12, 1901, the Hart-Parr Company of Charles City, Iowa, was organized under the laws of the state of Iowa with a capital stock of $100,000. It

An interior photo of the Hart-Parr Works shows the assembly of 22-45 Hart-Parr engines. The serial numbers of the engines were written on the side of the blocks during assembly. These serial numbers were also the serial numbers used on the tractors. Notice the completed tractor sitting in the rear of the engine assembly room. *Author's collection*

broke ground on its new 3-acre building site on July 5, 1901. By Christmas, the transfer of inventory and operations of the Wisconsin business was completed, and the new plant was up and running with fifteen employees.

With the new plant up and building gasoline engines, Hart and Parr started to design a gas engine that could drive itself to the field. Since gas engines were used for belt work, an engine was needed to run the thresher and other implements. It would be much more useful if it was mobile. The winter of 1901–02 was therefore dedicated to building the first self-propelled gasoline traction engine.

That first traction engine was known as Hart-Parr No. 1. It was sold and delivered to David Jennings, a farmer west of Mason City, Iowa, in July 1901. One of the Hart-Parr mechanics proudly drove the engine 36 miles to the Jennings farm while Hart and Parr followed. While the engine was crossing a rickety wooden bridge, the bridge collapsed and dropped the engine and driver into the muddy creek below. A team of horses pulled the engine from the creek; it was then cleaned up and driven the rest of the way to the Jennings farm. Despite its precarious start, this machine operated successfully for seventeen years before it was retired.

A new and improved engine model was built in 1902, and fifteen traction engines were built in the next year. Hart and Parr felt they now had a successful operating model, so they put on their business caps to figure out how to build it in the most efficient manner. With parts standardization and mass production in mind, Hart and Parr began to focus everything on the booming tractor industry. At this point, the word "tractor" was not yet commonly used. It was in 1906 that the Hart-Parr sales manager combined the words "traction motor" and began using the word "tractor" in all future advertising.

In 1905, the Hart-Parr Company discontinued production of stationary gas engines to put all its effort into the tractor industry. It was a big step, but it gained the company the title "Founders of the Tractor Industry." That same year, the company started an expansion program that would extend the

The Hart-Parr Company was constantly in some stage of construction. The extension of the erection shop, gray iron foundry, steel foundry, and the powerhouse is shown here in this photo from 1907. *Author's collection*

This artist rendition shows the massive expansion by Hart-Parr. The photo was taken in 1928, a year prior to the merger between Oliver, Hart-Parr, and Nichols & Shepard. *Author's collection*

original buildings and expand the foundry, machine shop, and shop building. In addition, the company purchased adjoining lots for future expansion.

Demand for the Hart-Parr tractors was so great that the plant had to be operated day and night. By the end of 1906, Hart-Parr had built nearly five hundred tractors, and expansion was once again at the forefront. A new erecting shop was built, and the powerhouse and foundries received additions the next year.

By 1910, Hart-Parr had a 70 percent market share of the tractor industry. The company was well-represented in both the export and domestic markets. The "Founders of the Tractor Industry" were admired and copied by all who wanted to compete. By 1912, the plant needed even further expansion. In order to raise working capital to make these additions, the company reorganized with $1 million common stock and $1.5 million preferred stock.

During World War I, most manufacturing plants were converted to manufacture items that assisted the war effort. While Hart-Parr manufactured many different items during this time, one of them was building the steel wheels for ammunition trucks. Nash subcontracted with Hart-Parr to manufacture the wheels for these trucks, and about eighteen thousand were produced in the Charles City plant. *Author's collection*

World War I was rough on the Hart-Parr Company. It had invested an enormous amount of money in tooling up for a British ammunition shell contract. While in full production, and with the order nearly completed, the contract was cancelled, causing the company a catastrophic financial blow. By 1917, company stock was virtually worthless, and the bankers holding the notes took over much of the decision making where finances were concerned. Hart was not happy with this arrangement and left the company.

The company received another blow when the drop forge building was destroyed by fire in October 1917. It was the largest fire in Charles City's history. The estimated loss of the building and contents was approximately $200,000. Although the company appeared to be down and out, about one thousand employees were still working at the Hart-Parr plant.

Tractor production once again resumed after World War I. During the spring of 1919, the plant turned out an average of twelve tractors per day. Then, with the new design of the smaller tractors, orders began to pour in. For every tractor built, there were ten unfilled orders. This success was short-lived, however, as the depression of the early 1920s took its hold on the country. In 1923, Parr left the company that he had helped found. A year later, the chief engineer, Erwin Frudden, also left the company and went to work for Allis-Chalmers.

By 1926, its twenty-fifth anniversary, the Hart-Parr Company had worked its way back up to full production. Parr returned to the company to work in the experimental department. The Fordson tractor was showing great dominance in the market, and Hart-Parr would struggle with this force for the next few years.

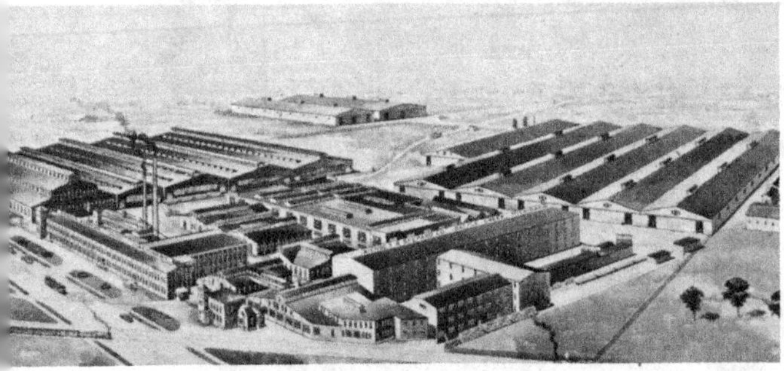

The Nichols & Shepard Company experienced continuous expansion at this plant, although the details are scarce. The new and larger buildings tower above the small original building where it all began in 1848.
Author's collection

Nichols & Shepard: Battle Creek, Michigan

John Nichols began his career in the farm equipment industry in 1848 when he opened up a blacksmith shop in Battle Creek, Michigan. He took David Shepard on as a partner in 1850, and together, they built a small line of farm machinery. In 1857, Nichols & Shepard developed a threshing machine, the Vibrator, which used vibrating straw racks instead of an endless belt. In the first year, ten machines were built and immediately sold. This was the beginning of a lucrative career for the team.

Due to the potential growth, the company was incorporated in 1866 under Michigan law. In the 1880s, Nichols & Shepard began to manufacture its own steam engine to pair up with its already-popular thresher. John Nichols passed away in 1891, and his son, E.C. Nichols, took over the operation of the company.

Nichols & Shepard introduced its famous line of Red River Specials threshing machines in 1900. The early machines, like most of that era, were made of wood. In 1915, the first all-steel machines were introduced. During this same time period, the company also introduced its first gasoline tractors, which it continued to produce into the 1920s. Lauson-built tractors were sold under the Nichols & Shepard name as smaller tractors began to replace the large prairie models. John T. Nichols, grandson of the founder, now ran the company, which soon came out with a combine and a corn picker. Nichols & Shepard turned its focus to manufacturing harvesting equipment, which became its lifeblood.

The Big Merger

The Great Depression of 1929 was around the corner, and big changes were in line for farm equipment companies. In October 1928, the Hart-Parr company amended its incorporation certificate to increase available stock. Sam White Sr. was elected as a new company director.

White was a Harvard graduate who had worked his way up to become the manager of the securities department for the Federal Reserve in Chicago. White strongly believed in the Shakespearian quote, "Neither a borrower nor lender be." It became his job to find companies in need and match them up with good partners in the form of mergers.

Familiar with the Oliver Chilled Plow Works, White knew the company needed to add a tractor to its product line; it had been experimenting with its own tractor for several years. White was also very familiar with Hart-Parr and its situation. The Nichols & Shepard Company was in need of additional products to add to its line as well.

White worked to put together a merger of these three companies. Together, they could supply the farmer with a tractor, combine, plow, or any tillage tool needed. J.D. Oliver was running the Oliver Chilled Plow Works and was getting up in years. He knew that his sons didn't have much interest in taking over the company and saw trouble ahead.

Any merger, however, begged the question: When three successful, long-standing companies are joined together, whose name gets to stay on the business? The combination of the three companies had a value of $50 million, but the assets of Oliver alone were more than the combination of Hart-Parr and Nichols & Shepard. J.D. Oliver therefore declared, "My name goes on the company or the deal is off." Thus, the Oliver Farm Equipment Company was born on April 1, 1929. Another interesting fact in this merger was that J.D. Oliver did not want stock in the company. He wanted his contribution to the new company to be paid off in cash. This move proved to be a wise decision because the stock market crash could have wiped out the Oliver fortune in later months.

Within two weeks, the American Seeding Company joined the Oliver Farm Equipment Company. The company was well known and established as

The $50 million men: these are the directors of the newly formed Oliver Farm Equipment Company. In the top row from left to right: John T. Nichols, vice chairman; Melvin W. Ellis, former Hart-Parr president; Walter A. Weed, former sales manager for the Oliver Chilled Plow Works. Seated in front are J.D. Oliver and James Oliver II.
Author's collection

a manufacturer of grain drills, planters, manure spreaders, and cultivators. It was located in Springfield, Ohio, and had assets of $3.5 million.

Oliver was now among the top four manufacturers in the farm equipment industry with six thousand employees and annual sales in excess of $30 million. Oliver opened its corporate office at 400 West Madison in Chicago. At this location, it was close to banks, lawyers, good transportation, the best advertising agencies, and anything a big business needed. It was conveniently located between each of the four divisions.

The Hart-Parr portion was tagged the tractor division. The South Bend plant became known as the tillage division. Battle Creek became the harvesting division, and the Springfield, Ohio, plant became Oliver's planter division.

That same year, the Oliver Farm Equipment Company acquired a short-line company out of LaCrosse, Wisconsin—the McKenzie Manufacturing Company—well known for its potato machinery. This was a familiar product to Oliver because the Chilled Plow Works had sold the McKenzie planters for many years through its agents.

With the country in a depression, the company was doing what it could to trim operating costs. One step was to close the McKenzie plant. The potato planter production line was transferred to South Bend.

The Oliver Farm Equipment Company managed to survive the Great Depression, a feat not shared by many smaller companies. Corporate reports showed that the shares at the end of 1943 were worth $5.26. This was an 8 percent increase in worth over the previous year's stock value. The company was moving ahead, and the next step was to add a missing equipment line to its portfolio.

Ann Arbor Baler: Shelbyville, IL

Although the Ann Arbor Machine Company originated in Ann Arbor, Michigan, it was located in Shelbyville, Illinois, when Oliver took interest in it. Horace Tallman was a commercial hay dealer from Shelbyville in the late 1800s. Not only did he sell hay, but he also sold the equipment to make hay, having acquired the selling rights to the Ann Arbor baler in 1899. This agreement made him the sole Ann Arbor dealer in the state of Illinois.

The Ann Arbor Hay Press Company was under financial stress by 1920. Tallman decided to purchase it, and in 1921, the company was moved to Shelbyville. It was at this location that Tallman and his sons developed the first pickup baler, an innovation that put the Ann Arbor baler on the forefront of the farming implement world.

The Tallman family operated their successful baler plant until Oliver approached them in 1943. The company was facing stiff competition, and its product line consisted of little more than balers and railroad tools. The Tallmans knew the company needed more products, which it wasn't set up for, or it needed to be sold. Oliver was the answer.

On November 1, 1943, the Tallman brothers retired and Oliver took over the plant, making Shelbyville

The American Seeding Company was formed from a merger of several companies. One of the plants that fell under the American Seeding flag was the Hoosier Drill Company. This plant, located in Richmond, Indiana, became part of American Seeding in 1903. However, International Harvester acquired the Richmond factory and its product line in 1920, prior to the Oliver merger. *Author's collection*

its hay division. Oliver purchased the inventory and the product line but leased the plant. A month after Oliver took over, the plant burned to the ground. This was a major blow to the city since the baler plant was its largest employer. The Tallmans rebuilt the plant for Oliver, however, and it was up and operating by July with the most modern equipment in the industry.

Cletrac: Cleveland, Ohio

The Cleveland Motor Plow Company was started by Rollin H. White in 1916 with the introduction of the Model R crawler. A year later the company's name was changed to the Cleveland Tractor Company (Cletrac). The company was based in Cleveland, Ohio, on Euclid Avenue. The entire focus of the company was the crawler line, which proved to be strong competition against Caterpillar, and White held many patents on a number of his innovations.

Like most of the other companies involved in the merger, White also brought his family into the business. Rollin's son, King, took over the company, and the torch was passed. By 1944, King was facing health problems, however, and had no family members interested in taking over the business. Oliver approached White about adding the crawler line to its equipment family, and on November 1, 1944, Cletrac became part of the newly formed Oliver Corporation.

The Cleveland plant continued operation as Oliver's crawler division with few changes in personnel, occupying over 100 acres of ground with 336,000 square feet under its roof. It had more than one thousand employees with sixty men working in the tool room to maintain dies, jigs, and tools. The factory held forty-four thousand different machinery and assembly operations with 340 feet of moving tractor assembly line, as well as 1.18 miles of material conveyed equipment.

A.B. Farquhar Company: York, Pennsylvania

Arthur Briggs Farquhar decided when he was nineteen years old that he wanted to become a millionaire. He took a train, ferry, and stagecoach to New York City and approached every millionaire he could find to ask for advice.

After returning home to York, Farquhar told his father that he was going to manufacture something mechanical. He went to work as an apprentice at an implement company and worked his way through the ranks. In 1887, York, Pennsylvania's greatest industrialist endeavor was incorporated as A.B. Farquhar Ltd. with a capital stock of $500,000.

Farquhar began manufacturing grain drills, plows, steam engines, and portable sawmills. Through the years, the company added conveyors, cider presses, and sprayers to its product line. Farquhar acquired the Bateman Manufacturing Company's Iron Age Potato Machinery Line in 1930.

By 1946, Farquhar had 1,200 employees on the payroll at the York plant. A few years later, in 1951, Oliver announced the acquisition of the A.B. Farquhar Company. A.B.'s son, Francis, stayed on as chairman of the board. Oliver saw the company as a supplement to its product line. The plant occupied eight acres and had 430,000 square feet of floor space. The acquisition gave Oliver a total of ten thousand employees.

In the 1940s, the Cleveland Tractor Company was the world's largest exclusive crawler manufacturer. It built no tools or products for the crawlers. The focus was 100 percent on the crawler industry. *Author's collection*

Battle Creek Aviation Division: Battle Creek, Michigan

Oliver's Battle Creek plant was home to the company's combines and corn pickers. Oliver, however, had been awarded a government contract to build the fuselage on the Boeing RB-47E aircraft. To fulfill the manufacturing requirements, Oliver purchased the land, buildings, and office equipment of the Goss Printing Press Company in Battle Creek at a cost of $1.68 million. This spacious plant, situated on 11.4 acres, became known as Battle Creek Plant No. 2, or the Aviation Division. The contract with the government lasted long enough to build 255 Boeing fuselages. Upon completion of this contract, Oliver acquired the Chris-Craft line of outboard motors, and the Battle Creek plant became the outboard motor division of Oliver.

The Be-Ge Company was in full swing manufacturing 76 mm aircraft gun barrels through a government contract when Oliver paid it a visit. This modern plant was very enticing to a company looking to expand in the industrial market. *Author's collection*

Be-Ge: Gilroy, California

The Be-Ge Company, based in Gilroy, California, began in 1935 when J.E. Bussert and Al Gurries joined forces. Together, they saw a need for machines that would build roads and dams and that would level land. To operate in the most efficient manner, this equipment would have to be fitted with hydraulics.

Be-Ge established itself as a company with reliable equipment that could easily do the job. The hydraulic units made the company popular in the international market for both power units and machinery.

In 1953, after two years of negotiations, Oliver purchased all of the outstanding capital stock of Be-Ge for $1.314 million. The plant was situated on 27 acres with approximately 229,000 square feet of floor space. The buildings were relatively new with construction completed in 1947.

Be-Ge continued to operate under its own management after the acquisition, and Be-Ge scrapers were coupled with Oliver's industrial line. This would be Oliver's final purchase before it, too, would be acquired by a larger company.

HART
PARR
HART-PARR

CHAPTER 2

AGRICULTURAL TRACTORS

At the turn of the twentieth century, the land was farmed with animal or steam power, both labor-intensive methods. With horses or oxen, the animals had to be tended to year-round. With the steam engine, farmers had the additional tasks of building a fire early in the morning before work could begin and hauling water for the running engine throughout the day.

The gasoline traction engine offered a way to free up many of the hours of labor previously put in on and off the field. When the engine was fired up in the morning, there was instant power. There was no need to haul water, and only one person was needed to operate a gasoline engine. Although regular maintenance and gas were required to keep the tractor running, feed and water were not.

Hart-Parr introduced its first gasoline traction engine in 1902. It was basically a stationary engine mounted on a frame with a chain drive from the crankshaft to the spur drive. Even though the tractor was capable of pulling implements around, it was not a "plow tractor."

The first Hart-Parr gas tractor engine, known as Model No. 1, was sold to David Jennings of Clear Lake, Iowa. It was driven 36 miles to his farm by Charles Hart and Charles Parr, a mechanic and a salesman. Before they reached the farm, the mechanic had to drive the tractor across a wooden bridge, which collapsed when the 19,000-pound machine traveled over it. Imagine the looks on the faces of Hart and Parr as they stared at their "baby" sitting in a muddy creek! The engine was pulled out of the creek bed by a team of horses, cleaned up, and delivered to the farmer as if nothing had happened.

Model No. 1 was only the beginning of a successful line of traction engines produced by Hart-Parr. Model No. 2 was built and sold the following year to P. Wendeloe of Ethan, South Dakota, for $1,800.

This Hart-Parr 17-30, built in 1903, is the oldest Hart-Parr tractor still in existence. Owned by the Smithsonian and restored in 2003, this tractor has spent the majority of its life being toured around by Oliver and on display in museums around the country. *Jeff Hackett*

The earlier row-crop model had the name Hart-Parr in larger letters than Oliver's name. Hart-Parr was a well-known industry leader, and Oliver wanted to capitalize on its fame and success by making the Hart-Parr name prominent when the model first came out. *Jeff Hackett*

This unit incorporated improvements from the previous model but still used the 17-30 engine.

In 1903, Hart-Parr built fourteen Model #3 units. This was the first successful run of a mass-produced tractor. One of the fourteen machines, now owned by the Smithsonian Institution of Washington, D.C., was originally sold to George Mitchell on August 5, 1903. This tractor, used as part of a threshing crew, has an interesting history. In 1924, Hart-Parr boasted the success of its tractors and tried to find its oldest tractor still in existence. A tractor with serial number 1207 was found. The company purchased the tractor back from George Mitchell for scrap price and toured it around the country as an advertisement to the success and durability of the Hart-Parr tractor. In 2004, this tractor was refurbished and a new canopy was built to celebrate its one hundredth birthday.

The word "tractor" was not even in use when the Model No. 3 was built. The term was coined in 1906 when Hart-Parr's sales manager, W.H. Williams, combined the words "traction" and "motor" to create the word "tractor." From this point on, "tractor" was used in Hart-Parr's advertising until it became a household term.

Twenty-four units of the 22-40 were built in 1903. These machines were built to be used in the fields and were equipped with a plow gear. This gear permitted the tractor to pull heavier loads with less chance of breakage of the drive gear.

The price of gasoline started to go up in 1905, and it was not a good selling tool for Hart-Parr's entry to the gasoline traction engine market. The company, therefore, soon developed a carburetor that was successful at burning kerosene, a cheap alternative to gasoline. All machines built in 1906 were equipped with the kerosene carburetor as standard equipment, and conversions were available for the earlier models.

One of the more popular models of Hart-Parr tractors is the 30-60. The model, built from 1911 to 1916, was nicknamed "Old Reliable," an indication of its success and reliability.

The Hart-Parr 12-24 was introduced in 1924 as a replacement for the 10-20 model. Advertised as a two-three plow tractor, it included a new steering system, enclosed gears, and a dry-plate clutch. It went through continual changes during production until it ended in 1930. *Jeff Hackett*

By the mid-1910s, the trend in tractors was turning toward smaller machines that could operate on the smaller farms. Hart-Parr's answer was a lightweight tractor built to pull a two-bottom plow. It was designed to be as simple as possible, and the number of parts was reduced in order to keep the tractor affordable for the general farmer.

This tractor was affectionately named the "Little Devil" and rated at 15-22 horsepower. It initially looked like a huge success, and the company had more orders than it could fill. Once the tractor hit the field, however, it experienced serious mechanical problems, and the model was recalled. Most of the tractors that were traded in for a substantial credit on a new model were destroyed. Of the 725 Little Devils built during production, only three are known to be in existence today.

With the demand still high for a small tractor, Hart-Parr went back to the drawing board and designed the NEW Hart-Parr. The first of these models was the 12-25. Minor changes were done on the first models, and the rating was soon increased to a 15-30. This model was the first Hart-Parr to be tested at the new Nebraska Test Laboratory.

These cross-motor-type tractors were quite successful for Hart-Parr and were built into the early 1930s. The independent PTO was introduced on the 12-24 and 18-36 models in 1928. A chain drive operated off the crankshaft and used its own clutch, which made it independent from the tractor.

While Hart-Parr sold its cross-motor models, the Oliver Chilled Plow Works was experimenting with a tractor of its own design. This tractor was built initially as a row-crop model and used a four-cylinder Hercules engine. A few of the later models were built in the standard tread variation. Approximately twenty-six of these tractors were built and tested from 1926–28.

When the merger between Oliver and Hart-Parr took place, Hart-Parr took the tractor to the Charles City, Iowa, plant, which was now the official tractor division of the Oliver Farm Equipment Company. The Oliver engineers studied the work done by the

The Hart-Parr cross-motor tractors were the first to use a water coolant radiator in order to cool the crankcase. Previous models used the large oil cooling towers that circulated the oil from the engine and cooled it with air. *Jeff Hackett*

Hart-Parr engineers and designed a completely new tractor. The new tractor would become known as the Hart-Parr Oliver Row Crop.

This new series of tractors eventually included the 18-27, the Row Crop model. The 18-28 was the same tractor but in a standard tread variation. The 28-44 was the biggest tractor in the lineup. All of these models were powered with a Waukesha engine.

In 1935, the company introduced a new six-cylinder tractor designed to run on 70-octane gasoline. The Continental-powered model was hence designated the 70. This was an all-new tractor with complete sheet metal, a hammock seat, and optional starter and lights.

The tractor division experienced somewhat of a turning point in 1937. By this time, both Hart and Parr had passed away. Any tractor built from here on out would only bear the Oliver name. The 70

The row crop tractor was first introduced in 1930 and it was the first tractor built by the newly formed Oliver Farm Equipment Company. After it was tested at Nebraska, the machine took on the rating of 18-27 but was still commonly referred to as the Row Crop. Initially, the Hart-Parr name appeared bigger on the radiator, but on the later models the roles were reversed and the Oliver name appeared larger. *Jeff Hackett*

Tractors originally equipped with steel wheels were often converted to use rubber tires in later years. This is an example of that situation, and the wheels are referred to as cut-offs since the cleats were cut off and replaced with a rim. *Super T*

The Hart-Parr Oliver 70 was the most modern tractor built in its time. Available with electric start and lights in 1935, it was the unstyled model that was replaced in 1937 with the streamlined model. It was the tractor that put Oliver back in the black following the depression. *Super T*

was updated with modern, streamlined styling. This tractor eventually became one of the most popular tractors built by the company, with more 70s built than any other model.

As one of the most successful Oliver models built, the 70 was built in several variations: row crop, high-clearance, standard, industrial, orchard, and Power-Pak. It was also built as a Cockshutt model. The styled version of the 70 was built from 1937 to 1948. *Author's collection*

The 18-27 and 18-28 were both designated Oliver 80 models. The 28-44 became the Model 90 and had a closed engine compartment similar to the 70. When this model was fitted with a high compression head, it was designated a Model 99.

In 1940, a little four-cylinder tractor was added to the Oliver family—the 60. It was powered with a Waukesha engine and was rated as a one-two plow tractor. It was offered in a row-crop and standard tread version.

One of the turning points in the success of the Oliver company came in 1947 when the Fleetline series was introduced. The design of these tractors started at the end of World War II. The company knew that there would be a post-war boom and an all-new line of tractors would be needed. Oliver originally approached Continental to provide the engines for the new series as it was already providing the engine for the 70. Continental said, however, that it would not be able to supply the quantity of engines requested by Oliver because it had just signed on to be a major supplier for Kaiser-Frazier automobiles. Thus, Oliver turned to Waukesha Motors, a company it had been working with since 1930. Together, they designed an engine to be used in the new series of tractors.

The 88 was the first of the Fleetlines to be introduced in mid-1947. It was an all-new three-four plow tractor powered by a 231.9-cubic-inch Waukesha engine. When first introduced, this tractor wore the same streamlined sheet metal as the 60 and 70, but it took on a look all its own in mid-1948. The 77 and 66 soon followed. These new Fleetline models became known as the "three beauties." They were heavily promoted, and the countryside was filled with billboards bragging about the modern features of the new Oliver models.

The 66 was the smallest in the lineup with a four-cylinder 129-cubic-inch Waukesha engine and was available as a row-crop, standard, or orchard model. With 22 horsepower on the drawbar, the 66 was rated a two-plow tractor. The LP engine was not available for this model.

The 77 used a six-cylinder Waukesha engine with 34 horsepower on the drawbar and a two-three plow rating. This was also available as a row-crop, standard, high-crop, or orchard model.

The single front-wheel models became available on the Model 70 and continued through the three-digit series. This was very popular for cultivating models and was used on all the vegetable specials. *Jeff Hackett*

The Oliver 80 Row Crop was an updated model of the 18-27. It was built from 1937 to 1948 and used a four-cylinder Waukesha engine. There were 136 Model 80s built with a Buda diesel engine, and only a few are known to still exist. *Super T*

The 1953 model 99 was built just prior to the introduction of the Super series. It used the back half of the old unstyled 99 and the front half of a new styled tractor using the Waukesha 195 engine. Both the gas and diesel models are pictured here. The tractor was built in the South Bend plant and rated as a four-five plow tractor. *Jeff Hackett*

A later entry into the Fleetline family consisted of the six-cylinder 99. This tractor was introduced in 1952 with a gas or diesel Waukesha engine and rated a four-five plow tractor. While the other models were built in the Charles City, Iowa, plant, the 99 was built in the South Bend, Indiana, plant. The 99 was available only as a standard tread model. It proved to be more suited as a plow tractor than anything. This tractor is often referred to as the 1953 99. It was somewhat of a transition tractor with a new front end but the four-speed back half of the older, unstyled 99. This tractor is one of the rare models of the Fleetline series as it was built for only a year prior to the introduction of the Super series.

The Super 88 used the popular Oliver/Waukesha 190 engine. Maximum drawbar horsepower was 47, while belt horsepower was 55. Bore was increased on the Super 88 to make it a full 4-plow tractor. It was built from 1954 to 1958 and was then replaced by the 880.
Author's collection

Oliver designed the Fleetline series with parts interchangeability in mind. Engineers wanted the tractors to share the same look and be able to share some of the same parts, too. This worked well for the dealers by reducing the number of parts to stock in their inventory. This was also the first series to offer a variety of engine options such as LP, gas, and diesel.

In 1954, the Fleetline series was "Super-sized." The new Super series took on a slightly different look with its open engine compartment. In addition to the pipe-mounting holes in the frame for implements, the tractors also had flat plates incorporated into the frame for bolt-on implements. The engines on all the models had an increase in bore and stroke, thus raising the horsepower rating.

The Fleetline series ended with four models in the family. By the end of the Super series, there were six. The Super 55 entered the picture when the Super series was introduced in mid-1954. It was Oliver's first utility-type tractor and was built to compete with Ford. It used the same four-cylinder Waukesha engine that was used in the Oliver Super 66 and was rated a two-three plow tractor.

The Super 55 had many modern features to set it ahead of other utility models. It used double disk brakes, an independent PTO, three-point hitch, a front-end PTO for operating front-drive accessories, and a six-speed transmission that incorporated a Super Low gear. This machine was compact, easy to operate, loaded with features, and designed for comfort and ease of use.

The Super 88 was an updated Fleetline series. The Wheatland model used a stationary front axle, splash panels, and fender extensions. When the Super series first came out, it had green wheels, but after January 1, 1957, the color changed back to red. *Jeff Hackett*

The Super 66 was the smallest row crop model in the series. The single front wheel makes this unit a little more unique and collectible. The Super 66 used the same four-cylinder engine as the Super 55 and was eventually replaced by the 660. *Super T*

The 1955 was powered by the Oliver/Waukesha 310 turbocharged engine and rated at 108 PTO horsepower. When introduced, the 1955 4WD listed for $13,496 and had a bare weight of 11,720 pounds. *Jeff Hackett*

The Super 55 was one of Oliver's most successful tractors. This basic model was built for nearly twenty years, although it was later known as the 550. It had green wheels when it was first introduced with the other Super tractors. The company switched back to red wheels due to the negative response it received from the customers. *Super T*

The horse in the family was the Super 99. The Super 99 was an updated Fleetline 99 with a new back half and a six-speed transmission. The new Super 99 was available with the 3:71 Detroit diesel engine. The company still offered the Waukesha gas or diesel version, but the Detroit-powered model proved to be much more powerful and desirable.

The smallest Super series tractor entered the market in 1957. It was an offset model tractor designed to compete with the Farmall Cub. Known as the Super 44, it used an F-140 Continental engine. This tractor was used primarily for cultivating. The offset operator platform gave a perfect view of the rows ahead. This tractor was originally designed in Oliver's Battle Creek, Michigan, plant, but as production began, it was transferred to South Bend's plant to be built next to the Super 99.

In 1958, the Super series was updated to become the three-digit series. All of the models were updated, and the color scheme was changed. The popular red,

Work on the Super 44 originally began in Battle Creek, Michigan, but was transferred to South Bend where the Super 99 was built. Only 775 of the Super 44s were built in 1957 and 1958. *Jeff Hackett*

The 440 was very popular with the tobacco farmers as a cultivating tractor. It used an F-140 Continental engine and was rated a two-plow tractor. Due to the closing of the South Bend plant, the 440 was manufactured in the Charles City, Iowa, plant. Only 700 units were built by the end of production in 1962. *Jeff Hackett*

General Motors provided the engine for Oliver's heaviest tractors. The GM 71 series engine was optional in the Super 99 and standard with the 990 and 995. The 990 gutsy model used large bull gears that protruded through the top of the gear case to make it a strong plow tractor or industrial model. *Jeff Hackett*

The early 550s used a metal horizontal bar grille. In order to update the style to match the four-digit models, a fiberglass checkerboard grille was incorporated into the 550 in 1963. *Author's collection*

yellow, and green color scheme was replaced with Clover White and Meadow Green. The hood and grille were updated to give a more sleek appearance. The 660 and 440 models broke the rules and retained the same Fleetline/Super-style grille and hood but did incorporate the new green and white paint scheme.

The biggest changes in the number designations occurred in the 900 series. The Super 99 was upgraded to the 990. If the 990 was equipped with the Waukesha engine, however, it was designated the 950. The 990 only used the 3:71 GM engine. If the 990 had a Lugmatic torque converter, the tractor was known as the 995. The 900 series tractors were available only as standard tread models. By the time these tractors were built, production had been transferred from the South Bend plant to the Charles City plant.

The smallest tractor in the three-digit lineup was the 440, merely a Super 44 updated with the new color scheme. The 550 utility model remained a very popular tractor and was built in large quantities. The series also included the 660, 770, and 880. In 1960, Oliver was ready to update its tractors again even though the previous series was only two years old. The trend in farming was turning to more acres and more efficiency. The 1800 and 1900 were introduced as the 100 series tractors. The 1800 was introduced with the 265-cubic-inch Waukesha gas engine or 283-cubic-inch diesel engine. It was rated a six-plow tractor and was referred to as the 1800A or the checkerboard model. The 1900 used the 4:53 Detroit diesel engine and was rated as an eight-plow tractor. The first two models were known as the checkerboards or A models and used a decal on the side for model number identification with a checkerboard background.

There was quite a gap between some of the models, but it wasn't feasible to tool up to build more tractors. Instead, Oliver turned to David Brown of England to act as a supplier. The 850 David Brown was painted green, and a grille was designed to match the current Oliver lineup. The tractor was decaled an Oliver 500 and rated at a two-three plow tractor with approximately 30 horsepower in both diesel and gas. The 500 was built from 1960–63.

In 1962, Oliver requested a slightly larger tractor to go along with the 500. David Brown took its Model 990 and styled it to look like the other Oliver model. It was decaled an Oliver 600 and rated as a three-four plow tractor. This partnership ended in 1963 when David Brown began to market its own tractors in the United States.

The White Motors Corporation acquired Oliver in 1960 after the introduction of the 1800 and 1900. In 1962, White Motors acquired the Cockshutt Company headquartered in Ontario, Canada. Oliver took the nameplates used on the Cockshutt tractors and put them on the Oliver tractors. These nameplates were used throughout the end of the Oliver tractors'

The Hydra-Hitch was an Oliver feature offered with the 770 beginning in early 1957. Kits were also available for the older Fleetline and Super series tractors. The Hydra-Hitch was operated by 3- or 4-inch Hydralectric cylinders and incorporated free hook-up lower links. *Jeff Hackett*

The 880 was the upgraded version of the Super 88. Introduced in 1958, gone was the green, yellow, and red combination, and new were the Meadow Green and Clover White colors. Those colors would remain with Oliver until the end. *Super T*

The 1800 was produced as the A, B, and C models. The A model was referred to as the checkerboard because of the decal on the side. When the B model came along in 1962, a nameplate replaced the decal and that plate is often referred to as the name spear. *Super T*

The mechanical front wheel assist was introduced on the 100 series tractors. The hydraulic front wheel assist came along with the next series and only lasted a few years. However, the old reliable mechanical system stayed on the scene throughout production. *Super T*

When the 100 series was introduced in the fall of 1959, a heavy cast iron grille was added to the tractor to compensate for proper weight distribution. This heavy grille was used from 1960 until the end of production in 1976. Fender fuel tanks were another option that helped the farmer keep the big tractors in the field for the full day. *Super T*

The 1900A model used the checkerboard grille like the 1800A. These were the only two models that used the decal. The 1900 used a GM 4:53 engine and was rated at 89 PTO horsepower. The GM-powered tractors are easily distinguished from a distance by the dual air cleaners. *Jeff Hackett*

production. At the same time the nameplates appeared, the 1800 and 1900 were upgraded and called the B Model. The 1800 engine was changed to a 283-cubic-inch gas engine and 310-cubic-inch diesel engine.

After the acquisition, Oliver also added the 1600 to the lineup. It was classified a four-five plow tractor and was available in gas, LP, or diesel. Drawbar horsepower observed at the 1963 Nebraska test was 48, and it was available with optional Hydra-Power drive.

The year 1963 brought about a new feature in the big tractor market. Four-wheel drive mechanical front ends were available for the 1800 and 1900 models. This was a modern, big-horse feature that set these models apart from the rest. The later series offered a hydraulic front-wheel assist built by Levy of Canada. This company sold its front ends as aftermarket accessories for many brands of tractors.

The next series to join the Oliver family was the 50 series and included the 1650, 1850, and 1950. This group was basically an updated 100 series. With the larger model tractors, Oliver kept the 550 and 770 model in production to fill in the needs for the small farmer.

The 1600 was a late addition to the 100 series tractors. It was introduced as "new for 1963" and was available as a row crop, Wheatland, and hi-crop model, and could be purchased with a gas, diesel, or LP engine. The 100 series were the first tractors to be made available with a ROPS. *Jeff Hackett*

The 1600A series used a 23-cubic-inch engine but was upgraded to a B model with a 265-cubic-inch engine. This model joined the 100 series late and was replaced by the 1650 in late 1964. When equipped with the Power-Adjust wheels, the 1600 was a versatile plowing tractor. *Super T*

Before the end of the 50 series, the 770 was replaced by the 1550. The 1750 was also added to fill in the gap between the current models. A later addition to the 50 series family was a larger, more powerful model—the Oliver 2050. It was powered by the White-Hercules six-cylinder diesel naturally aspirated engine. The 2150 was the same basic model but used a larger clutch to handle the additional power of the turbocharged variation. These models were built only from 1968–69. Less than four hundred of the 2050 models were built, while just over one thousand of the 2150s were produced.

To fill the need in the small tractor arena, Oliver contracted with Fiat of Italy to provide its tractors for

The 770 was often coupled with the 778 loader. The standard bucket had a 3/4-yard capacity, but a snow and coal bucket with a 1 1/4-yard rating was optional. The lifting capacity of this unit was 5,100 pounds. *Author's collection*

Prior to using the Oliver 770 as their power unit, Harrison built a similar machine using the Case 520. However, the 770 used a torque converter and revers-o-matic, which made it a much more versatile machine with a top speed of 17.3 miles per hour. *Author's collection*

Oliver began selling the Fiat model of tractors in 1965 and worked with Fiat until 1979. In the later years, the Fiat models were silver to match the new White Motors models. The last green Oliver Fiats were actually built in 1975. *Author's collection*

the company. The models available in the 50 series were the 1250, 1350, and 1450.

Another addition to the 50 series was the 1950-T. This tractor used the Waukesha 310 engine with a turbocharger. It was rated at 90 horsepower and replaced the more expensive, GM-powered 1950. This was Oliver's first turbocharged model and the first Oliver tractor to use the Over/Under Hydraul transmission and just over two thousand tractors were built.

In late 1969, Oliver introduced the 55 series as "New for 1970." The Fiat models were upgraded to the 1255 and the 1355. The rest of the models followed suit from the 1555 through 1955. Minor cosmetic changes set the 50 series apart from the 55 series. On the 55 series, the headlights were moved to the inside of the grille, and stripes were added to the fenders and decals. The dash was changed to wood grain to give it a more stylish look, and the tilt/telescope steering wheel was fitted with a rubber boot to keep the dirt out. For better operator comfort, the driver's platform used rubber bushings to help cushion the ride rather than the previously used rigid mounting.

Another addition to the 55 series was the 2455 and 2655 articulating four-wheel-drive models. These two tractors were actually Minneapolis-Moline tractors dressed in the green and white Oliver colors. They were the first models built by Minneapolis-Moline for Oliver. When the tractors were sold as the Minneapolis-Moline version, they were known as the A4T-1400 and A4T-1600. The 2455 was powered by the 504-cubic-inch diesel Minneapolis-Moline engine, while the 2655 diesel used the 585 Minneapolis-Moline engine and the 504-cubic-inch LP engine. These models used a two-speed drop box to multiply the gears of the five-speed transmission.

The 55 series tractors were first introduced in the fall of 1969. They are most easily identified by the location of the headlights, which were placed inside the grille. *Jeff Hackett*

The early 1855 models were plagued with engine problems caused by a lack of lubrication. This was an expensive fix for the company, but it was able to remedy the problem. Because of this, the 310 engine inherited a bad reputation in the 1855. *Jeff Hackett*

The 2255 was the only Oliver produced with a Caterpillar engine. Early models used the 3150 engine with production of 1,340 units. The later models used the popular 3208 and 820 were built with that configuration. The 2255 was built from 1972 to 1976. *Jeff Hackett*

The 2270 was essentially a Minneapolis-Moline G1355. When marketed in parts of Canada, these tractors were painted red and branded White. When they were sold in the US, they were designated as G1355 models. *Jeff Hackett*

In 1971, White Motors decided that selling Minneapolis-Moline tractors painted in Oliver colors would help with sales numbers and offer more variety to the Oliver dealer. It didn't really work, but the result was three more models added to the Oliver line: the 1865, 2055, and 2155. The models—the G950, G1050, and G1350—were current production models built in the Minneapolis-Moline factory.

The 2255 was added to the 55 series in 1972. This model used a 3150 Caterpillar engine with a 573 cubic-inch-displacement. In 1974, the engine was changed to a 3208 with 636-cubic-inch version in an effort to supply more reliability and power. The 2255 proved to be a very popular model in the Oliver line even though it was more expensive than most. When equipped with front-wheel assist and a cab or duals, this tractor gave the appearance of a massive machine. It was the only Oliver model to incorporate the use of the V-8 engine. It was also the last Oliver tractor to roll off the assembly line when production ended in 1976. From that point on, all tractors were silver and wore the White nametag.

CHAPTER 3

INDUSTRIAL TRACTORS

The first successful mass production of gasoline tractors took place in 1903. It wasn't long before the automotive industry followed suit, making the rutted cow paths and trails that served as roads insufficient for the rapidly increasing auto population. The growing auto industry had created a new market for the tractor industry by creating a demand for better roads.

Teams of horses and road graders couldn't keep up with the amount of work that had to be carried out in order to create more infrastructure. Tractors suited for construction and roadwork were needed to build roads where none had been before. It was this use that boosted the market for an industrial tractor.

Hart-Parr tractors first began performing roadwork in 1906. In the future years, there would be thousands of tractors converting the muddy trails into well-graded roads. The first roads in many western states were graded entirely by Hart-Parr tractors. At one time, more than three hundred Hart-Parrs worked on the roads in Iowa alone.

Although the Hart-Parr 30-60 and 12-27 were quite popular for roadwork, it wasn't until 1919 that the company actually designated a tractor for industrial use. The 18-35, which was built from 1915-18, was known as the "Oil King." It was used heavily in the United States and Europe and served well as an all-around workhorse. The wheel it used was cast together with the lugs. This proved to be a stronger wheel that eliminated the chance of the lugs peeling off under a heavy load.

Improvements to the 18-35 were made in 1919; this tractor was designated as the "35 Road King." Like most industrial models in the Hart-Parr line, this tractor used the belt horsepower rate as the numerical designation. It was made in very limited quantities with only about fifty produced.

The large, cumbersome tractors were on their way out and were replaced by the smaller models capable of the same amount of work. In 1920, Hart-Parr introduced the Hart-Parr Special Road-Maintenance tractor to

The Hart-Parr 18-35 Road King had a 10-inch bore with a single cylinder. The vertical cylinder operated at 500 rpm, and the oil cooler was located in the middle of the tractor on this model. *Author's collection*

The Oliver Hart-Parr Model 28 power unit was the smaller model of the two available during the early 1930s. This was an 18-horsepower model, but 28 was the belt horsepower rating. In early industrial models, they always used the belt horsepower rating. *Author's collection*

replace the Road King. The Road-Maintenance was based on the Hart-Parr 40. It used heavy cast wheels on the front and rear that were built with shorter lugs cast into the wheel so as not to damage the road. The tractor was fueled by kerosene to allow townships and businesses to operate the machine with an affordable and accessible fuel.

Later Hart-Parr models such as the 16-30, 12-24, 18-36, and 28-50 used solid rubber tires to make them more adaptable for industrial use. These wheels did not have the rubber cleats that we are accustomed to today. They were grooved with a diamond pattern or treads like a trailer tire. The tires worked well on hard surfaces, but they didn't do well in mud or field conditions.

In 1929, Hart-Parr merged with Oliver to create the Hart-Parr Oliver tractor. Together, the company created a new line of tractors, and the industrial versions of these tractors went into production in 1931. The serial numbers for the Industrials began with the 900000 number block. The first Industrial model was designated the Model 18, which was basically an 18-28. The Model 18 was produced for only one year before it was renamed Model 28. The 28 was produced until mid-1939 with 107 total units manufactured. In mid-1939, the 28 became the 35 Industrial. This tractor was basically the same as the 80 Standard, and only 328 of them had been built by the time production ended in 1945.

In addition to the industrial 18-28 and 28-44, the company also offered a unit known as a 28 Road Grader Power Unit and a 44 Road Grader Power Unit. This configuration was based on the current tractor models but with the front axle independent from the tractor and heavy cast dual wheels with balloon tires. The tires used a short lug running from one side of the tire to the other. With a setup like this, the customer could build his own road grader using the accessories produced by other companies such as Hawkeye or Duplex.

Hart-Parr Oliver also offered a basic power unit. Built in sizes 28 and 44, it was identical in design to the tractor but stripped down. There was no front axle or rear wheels; hence, no need for a steering column. Selling just the bare power unit meant customers could use it for any application of their choice. The power units were often set in road rollers or used as replacement units. They could be purchased in a variety of configurations and could be supplied with a front axle and left-hand or right-hand steering if desired. This allowed the unit to be mounted backwards in a machine, depending on the application.

When the Oliver 80 and 90 agricultural models were in production, the company sold them as industrial versions—35 and 50. This designation followed the earlier models and was the belt horsepower rating of the ag models. These models were also built as a road grader power unit or just as a bare power unit.

Oliver had a close working relationship with Ware Machine Works, which mounted many of its products on Oliver equipment as original equipment. The Ware Hydro-Trencher was a popular addition and transformed a tractor into a versatile piece of industrial equipment. The slab weights added 3,000 pounds to this machine. Trusses mounted under the rear axle helped support extra weight on the industrial versions of the Super series. *Jeff Hackett*

In 1945, the name was changed at serial number 900441 to the Industrial 80, and no major changes were made to the tractor. The one change that was made that sets this tractor apart from the 35 is the addition of side panels. The Industrial 80s were built in two variations: the 80A and 80B.

The 80A is very similar to the standard 80. Minor differences involved the transmission gear combination, rear wheel brakes, industrial color scheme, and equipment mounting pads. Early tractors in this group were equipped with the same worm and gear steering mechanism as the standard 80.

The second group of Industrial 80s began at number 900583 and was known as the 80B. They had the same engine and front main frame as the 80A but were equipped with the same front axle assembly as the 99. The 80 was also equipped with a heavier steering gear supplied by Gemmer and had 7.00 x 20 front tires. The rear tread could be either 54 or 62 inches. The front wheel tread on the 80B was 47-5/8 inches, just 1/8 inch wider than the 80A. The approximate shipping weight of the 80B was 6,200 pounds, 1,300 pounds heavier than its little brother. The A and B could easily be distinguished from each other when compared side-by-side.

Industrial 80 models were visibly different from the agricultural models even apart from the serial tags. The Industrial 80s were equipped with a spring-type drawbar, hood side shields, exhaust muffler, radiator curtain, upholstered seat, industrial rear wheel guards, three lights, and electric start with a battery and generator. The rear main frame was modified to provide a flat platform for the operator.

The 80 Industrial was built from 1945 to 1947. Only 684 of these units were produced. The Industrial version of the 80 would be equipped with engine side panels, while the ag version had an open engine compartment.

The Industrial 80 was not furnished with differential brakes. The tractor had foot-operated, internal expanding, mechanical wheel brakes. Brake drums were secured to each wheel with cap screws, and the brakes were applied by a single foot pedal mechanism. No mechanism was provided for braking one rear wheel at a time as a means of steering the tractor. Differential brakes could be installed on the 80

The 80 Industrial was built from 1945 to 1947. Only 684 of these units were produced. The Industrial version of 80 was equipped with engine side panels, while the ag version had an open side compartment. *Jeff Hackett*

Industrial at the dealership if the RC 80 rear main frame, differential, and final drive were installed. This was a very costly conversion, however, and was not considered practical. Industrial 80 tractors were equipped with a hand parking brake that operated on the bull pinion and simultaneously braked both wheels.

Another item that set the industrial model apart from the agricultural model was the method by which the headlights were mounted. On the ag model, the headlights were mounted to a bracket fastened directly to the side of the radiator. This set the front of the headlights flush with the front of the radiator. On the industrial model, a flat plate was used to set the headlights out in front of the radiator rather than flush with it. This practice started at number 900441, which was the beginning of the true "80" industrial designation.

Up until this point, the colors on the industrial models remained the same as those on the ag models. Oliver switched the colors on the industrials to orange during the late 1930s when it was producing the 60, 70, and 80. The only exception was the Oliver Airport 25.

The Airport 25 was added during the time period in which the Industrial 70 was being produced. Several original models have been found wearing green paint and a 70 serial number plate but decaled as a 25. To complicate things further, the 25 serial numbers fall into the same numbering sequence as the Industrial 70. Select pieces of Oliver 25 literature clearly picture a Model 70 tractor.

Most Airport 25 models were painted red with a white stripe down the frame. They had a drawbar on the front for pushing planes and an automatic coupler incorporated on the rear. This allowed the operator to effectively back up to a tow bar because the coupler automatically closed after contact was made, and the driver was hooked to the plane without having to get off the seat. This model also had an option snatch block or pulley to aid in moving planes around in small areas such as the hangers. Standard equipment on the Airport 25 was a foot gas pedal and enclosed platform fenders. This model was not equipped with a PTO because most of its work was limited to pushing or pulling.

Although the 25 was not technically an industrial model, Oliver marketed it to the airport managers as such. Sales literature showed all the duties it was capable of including plowing snow, sweeping, and mowing in and around the runways. An Airport 25 that is designated as such by its serial tag is a real gem in any collection today.

The industrial number designation changed on all models in 1945. From this point on, the number on the side of the industrial tractor was the same as on the ag tractor. The Oliver 50 was now the Industrial Oliver 99, and the Oliver 35 became the Industrial 80.

The smallest industrial model during the 1940s was the 60. This was a handy little four-cylinder tractor and was built from 1946–48. It was replaced by the 66 Industrial in 1949.

A model specifically designed for industrial work only was introduced in 1946. This was the Oliver 900 Industrial, built using the chassis and power unit of the 99. The fuel tank was relocated to the rear of the tractor, and the operator was moved forward for better visibility. The tractor featured hydraulically operated brakes, a heavy industrial drawbar with lateral or vertical adjustments, and a wide, bench-type seat with cushioned armrests. Rated at 63 horsepower, it was the largest wheeled tractor in Oliver's lineup at that time. When the Fleetline tractors went into production, the 900 Industrial continued to be built. The last one to be produced was in 1950, although they were sold for a few more years from inventory.

In the late 1940s, Oliver initiated a policy to differentiate between agricultural and industrial dealers. Ag dealers were permitted to distribute Oliver products for ag, and industrial dealers were permitted to distribute Oliver products for industrial uses, respectively, within the continental United

Weighing in at 2,375 pounds, the 66 was the smallest in the Fleetline Industrial series. It was most popular with road crews when fitted with a side mount mower. The 66 produced 26 horsepower. *Jeff Hackett*

Oliver offered many forklift variations, beginning with the Super 55 model. Mast heights varied from a lift height of 8 to 21 feet. Most forklifts were equipped with a shuttle-shift transmission for quick directional changes. Depending on the configuration, lift capacity was from 4,000 to 5,000 pounds. *Jeff Hackett*

States and Canada. Dealers were required to honor this policy and not cross over into boundaries set forth by the opposing division. When the situation presented itself, dealers were to refer customers to the other division. This same policy was in effect regarding allied implements. If the implement was to be used primarily for ag use, it had to be purchased from an ag dealer and vice versa. Oliver was very strict about this policy and encouraged both divisions to work together.

The 88 was introduced wearing the streamlined sheet metal of the 70 in 1947 and was built as an Industrial. Production of this model lasted into 1948 when the Fleetline sheet metal was introduced. The 88 and 77 both appeared as industrial models in 1948 and were followed the next year by the 66.

The Fleetline Industrial models differed from the ag models in color and in the fact that mounting pads were located on the frame. Also, the transmission used different gears for higher travel speeds. Unlike the ag models, the 77 and 88 used a Gemmer steering gear, the front wheel bolster was modified to allow for use of front-mount implements, and a heavier clutch was used.

The Super 77 and Super 88 continued to represent Oliver in the industrial market when the Super series was introduced to the ag market. The Super 66 was not available as an industrial model but was replaced by the Super 55 Industrial. This handy little utility tractor used the same engine as the 66 and could handle any task formerly done with the 66 and then some.

Oliver also produced a successful line of forklifts. These were based on the Super 55 and used a variety of manufacturers to provide the masts. The tractors were reversed, as was the transmission, so they could be driven backwards and still have a wide variety of gears. Model designations on these units were the 550, 551, 552, and in later years, the 2-44FL.

On the opposite end of the spectrum, Oliver introduced the Super 99 GM in 1955. This was truly

The Oliver 900 had the appearance of a massive industrial tractor. Powered by a 443-cubic-inch four-cylinder engine, it weighed in at nearly 13,000 pounds when fitted with dual rear wheels. A feature unique to this machine was an upholstered two-man seat. *Super T*

the big horse in the series with 72 horsepower on the drawbar and 80 on the belt. This tractor was the only industrial model out of the Super series that was not built in the Charles City tractor plant. The Super 99, like the 90 and 99 models before it, were built in the South Bend, Indiana, plant.

Shortly before production ended for the Super 99, a new model known as the Super 99 GMTC was introduced. The difference between the Super 99 and the Super 99 GMTC was the Allison torque converter that the new model was coupled with, an innovation that made it well equipped for pulling the Be-Ge scraper. The good relationship between Oliver and Be-Ge was evident by the compatibility of the two companies' models.

The three-digit series that was introduced next included the 770, the 880, and the 550. Production of the 900 series, not to be confused with the 900 Model, was transferred out of South Bend and moved to the Charles City plant. The 900 series had grown to three models by this time. The weakest of those was the 950, which used the Waukesha Model 195 engine. The 990 and the 995 both used the GM 3:71 engine. The latter, however, had a torque converter like the Super 99 GMTC.

One of the largest and most unique of the industrial models was the 990 scraper unit. In 1956, Jolly K. Harrison and Ben Harrison, sons of Gaines W. Harrison, devised a method of combining a tractor and scraper unit together without the use of the tractor's front axle. Harrison applied for a patent for this device on August 2, 1956. The Harrisons began using a Super 99 tractor for their combination vehicle. Evidently, Oliver saw this as an expansion to its contractor's line. The company immediately teamed up with the Harrisons to produce what is known today as the 990 Scraper.

Although the first scraper was known as a 990, it did not begin with a 990 tractor. The first 990 Scrapers were built using the Super 99 power unit.

Oliver's 995 used the GM engine coupled to the Allison torque converter, making it the beefiest tractor in the industrial line. Typically, a unit of this size was used for heavy dirt work pulling a scraper or a packer. *Super T*

The addition of a zero behind the 99 originally indicated industrial. The first 990 Scraper was built in late 1956, but it wasn't until 1958 that the three-digit series tractors went into production. When the tractor used in combination actually was the flat-nose 990, the same 990 designation was retained.

Although Harrison was already building Oliver scrapers, an agreement was signed in March 1957 between Harrison Hydraulics and Oliver. Harrison was an industrial distributor for Oliver equipment, so it's possible that early models were built by Harrison to be sold on its lot. When the pairing of the scraper and tractor proved to be a success, Oliver entered into an exclusive agreement with Harrison to put the Oliver tractor and Be-Ge pan together as one unit. The hitch to attach these two units and which allowed them to steer without the use of the front axle was built exclusively by Harrison.

The first order to be placed after the agreement was signed was for ten hitches to be delivered by April 15, 1957. Super 99 tractors were shipped via rail from South Bend minus the front axle. These were known as two-wheel special tractors. The scrapers were shipped in from Be-Ge. The tractor, hitch, and scraper were all put together in Columbia, South Carolina, and then shipped back to Oliver.

Harrison built three basic models of Oliver scrapers. Early models like this one were powered by a Super 99 tractor. The next models used the 990 or 995 tractors. This particular machine was an early prototype model. *Author's collection*

Another 30 units were ordered next and were to be delivered by July 31, 1957. Harrison warrantied

The early 990 scrapers used red wheels, but later models were painted all yellow. The GM-powered machine had a 6.7-yard capacity when the pan was full and weighed in at nearly 18,000 pounds when empty. The retail price of the 990 scraper was $16,995 in 1958. *Jeff Hackett*

the hitches and modifications to be free of defect in material and workmanship during the first 1,000 hours of use or first six months after sale to the user, whichever came first. Harrison agreed to replace parts if any defects in material or workmanship in the hitch or the installation work developed.

Harrison granted Oliver the exclusive right to manufacture hitches and installations for self-propelled scraper units according to Harrison's designs. Harrison also granted that Oliver was the sole owner of all rights to such hitch and installation designs, and Harrison was not to manufacture self-propelled scrapers or hitches for anyone besides Oliver prior to December 31, 1973. The patent that had been applied for back in 1956 was also to be assigned to Oliver, and if Harrison obtained any future patents on hitch or installation design, they were to be assigned to Oliver, as well.

Harrison's patent reads:

"This invention relates to scrapers of the self-propelled type used in road and other grading work, which comprise a tractor-like power unit and a scraper apparatus towed by the power unit. The power unit is supported on one axle and the scraper on another, and the two are joined to provide a single machine.

"Many of these scrapers and the prime mover or tractor therefore are designed as a single unit and are especially adapted solely for the purpose of scraping. They are extremely costly, and the tractors are not suitable for any other purpose. As a result, the initial cost of these units has limited their use to larger contractors, and further when there is no use for the equipment, the prime mover often stands idle.

"Accordingly, one of the objects of the instant invention is to provide a tractor-scraper combination embodying a power unit or tractor, which can be used for other purposes when the scraper is not needed. In addition, the invention has as an object, the reduction of the cost of such equipment.

"Also, another object of the invention is to provide a tractor-scraper unit in which an ordinary tractor can be used as a motor-powered means for the scraper unit with minor modifications."

Perhaps the patent was one of the selling points that interested Oliver in the Harrison unit. The sales literature promoted the use of the tractor unit to operate independently of the pan with simple modifications. Although they are supposed to be minor, the modifications did involve bolting the front axle back under the tractor, unhooking quite a few hydraulic lines, and converting the steering from hydraulic back to the direct connection to the front axle.

The 990 Scraper was offered with a torque converter from the very beginning. If it had been a tractor, the 990 would have been designated Super 99 GMTC. All of the scraper units, whether they were powered by a Super 99, a 990, or a 995, were all known as 990 Scrapers. Harrison continued to build scrapers into production of the 990 and 995 tractors.

The last scrapers available, most likely leftovers from 1960 production, were sold in 1962. White

Oliver offered the Strait-line loader for the Fleetline and Super series tractors, which was advertised as the most versatile loader on wheels. The unique loader was built by Maine Steel and could be operated from either direction. From one location, the tractor could dip into the pile, and by pulling forward just a short distance, the bucket could be tipped over the top of the machine to load the truck on the other side. While it may seem clumsy by today's standards, it is a unique machine to see in operation.
Jeff Hackett

Motors did not take Be-Ge when it acquired select assets of Oliver in 1960; therefore, it no longer owned the company that had manufactured the pan portion of the scraper.

Harrison had also been acquiring Oliver tractors on an OEM basis for its own manufacture. The Harrison Loader, sold by Harrison, was based on the Oliver 770. At a later date, when Harrison based the unit off the Oliver 1550 and 1650 machines, Oliver did offer them as part of the Mighty Tow line as the White Log Loader.

Oliver had been successful in the industrial tractor business since the 1930s, and both ag and industrial divisions operated under the same flag. When White Motors stepped into the picture, however, the two companies were separated and the Oliver Industrial Division was created.

The Industrial Equipment Manufacturers Council, a national organization, was formed in the early 1960s. Its function was to provide industry data such as sales numbers, trends, and general information to industrial manufacturers. In 1966, Sam White announced the formation of a new division for Oliver's industrial line. All industrial models would fall under the flag of the newly formed Contractors Equipment Division (CED).

There was great potential in the construction business at that time, and this was Oliver's way of reaching out to the growing market. The new series was demonstrated at a traveling road show billed as "Industrial Progress in Action" in April 1966. The first stop was in Charlotte, North Carolina. From there, the show traveled to Harrisburg, Pennsylvania, and finally to the last stop in Topeka, Kansas.

A letter from the Oliver Corporation stated that effective January 1, 1967, all equipment would receive a new number designation and new name. Thus, the Mighty-Tow, Mighty-Hoe, and Mighty-Lift series was born.

The smallest tractor in the new CED line was the 2-44, the industrial version of the 550. The first number represented two-wheel drive, and the second set of numbers designated the horsepower rating. The following list shows the industrial number compared with the agricultural version on which they are based.

The Oliver-White 4-115 was the industrial version of the Oliver 1950. It was powered by a 4:53 GM engine and was one of the most popular of the Mighty Tow series. The Allis-Chalmers D-21, John Deere 760, and the IH 21256 were the competition for the 4-115. *Jeff Hackett*

The 2-144 Mighty Tow was the equivalent of the Oliver 2150. It was powered by a 478-cubic-inch White-Hercules engine. This model was the largest offered in the Mighty Tow series. When equipped with front and rear planetaries, this model is known as the 4-144 EHD (Extra Heavy Duty). *Jeff Hackett*

2-44	550 2-WD
2-62	1550 2-WD (cast frame)
2-63	1555 2-WD (steel frame)
2-78	1650 2-WD
4-78	1650 4-WD
2-105	1850 2-WD
4-105	1850 4-WD
2-115	1950 2-WD
4-115	1950 4-WD
2-128	2050 2-WD
4-128	2050 4-WD
2-144	2150 2-WD
4-144	2150 4-WD
4-144 EHD	2150 4-WD with planetaries
2-44 MHS	550 Mobile Home Special
2-78-17	1650 2-WD w/1678 loader & 1617 backhoe
4-78-17	1650 4-WD w/1678 loader & 1617 backhoe
2-78-L	1650 2-WD with 1678 loader
4-78-L	1650 4-WD with 1678 loader
2-62-L	1550 2-WD with 1678 loader
2-62-15	1550 2-WD w/1678 loader & 1615 backhoe
2-78-15	1650 2-WD w/1678 loader & 1615 backhoe
4-78-15	1650 4-WD w/1678 loader & 1615 backhoe
2-44-L	550 w/578 loader
2-44-LL	550 w/588 loader
2-44-13	550 w/578 loader & 613 backhoe
2-44-13LL	550 w/588 loader & 613 backhoe
2-44-FL	552 forklift

Later additions:

2/4-80	1655 (steel frame)
2-80-17	1655 & 1617 backhoe
4-80-17	1655 & 1617 backhoe
2-63FL	1555 forklift (steel frame)
4-78FL	1650 4-WD forklift
2-78-LL	1650 2-WD log loader (Harrison)
4-78-LL	1650 4-WD log loader (Harrison)
4-25-L	skidsteer loader (Erickson)
4-30-L	skidsteer loader (Erickson)
4-62-Hi-Lift	Hi-lift loader (Lull)
4-95-Hi-Lift	Hi-Lift loader (Lull)
4-116-Hi-Lift	Hi-Lift loader (Lull)
4-50-AL	articulating loader (built by Dynamic Industries)

Based on the Oliver 1650, the 2-78-LL and 4-78-LL were introduced as part of the Contractors Equipment Division. List price on the 4-78-LL was $18,666. Oliver/White ceased sales of the log loader after 1970, but Harrison continued to sell it for a few more years as the Harrison Loader. *Author's collection*

Many changes were taking place in the White Motors family by the end of the 1960s. Both Oliver and Minneapolis-Moline operated under White, including approximately thirteen hundred Oliver dealers and nine hundred Minneapolis-Moline dealers in 1969. The volume of Oliver's production and sales was twice that of Minneapolis-Moline. When White decided to lump the two companies together, people and equipment were jumbled everywhere. The announcement that the CED would be formed anew as White Construction Equipment (WCE) was made in the fall of 1969, less than two years after the original formation of the CED.

The Contractors Equipment Division operated as a division of the Oliver Corporation, while the White Contractors Equipment Division operated as a division of White Motors Corporation. The president of White Contractors was Jim Wormley,

The 2-44 Mobile Home Special was based on the Oliver 550 with 44 horsepower. Marketed to mobile home park operators and mobile home manufacturers, the 2-44 incorporated a special hitch that made hooking up to and moving mobile homes a one-person job. The front of the tractor incorporated the heavy weights that were also used on the forklift units, assuring the front end could stay on the ground. *Author's collection*

Near the end of the White Construction Equipment line, equipment was painted Euclid green to go along with the Euclid line that was purchased by White. The industrial tractor line was discontinued shortly after the color change. *Author's collection*

The Tampo Company used many of Oliver's Power-Paks in its roller line. This Super 66 model with the reverse-o-torc transmission made it a popular unit in the construction field. *Jeff Hackett*

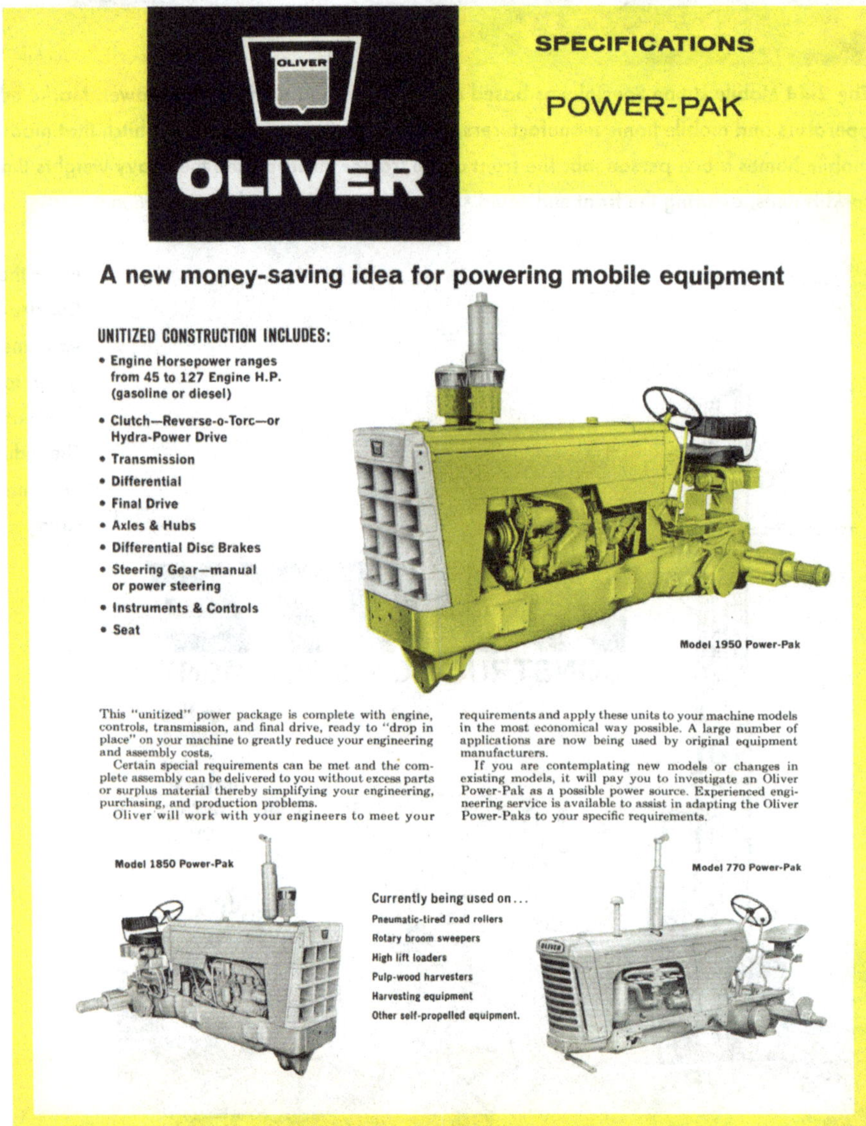

Power units that became available in the early 1930s were still in production in the late 1960s. These Power-Paks were quite popular with companies building industrial equipment. Many Oliver-powered road rollers are still in use today, which is a testament to their durability. *Author's collection*

At first glance, one would hardly know that this was an Oliver-powered unit. However, Hancock built several different models using an Oliver Power-Pak. This particular model is using Oliver's Waukesha 310 engine. Oliver's clutch, transmission, and dash were also used. *Jeff Hackett*

president of the Oliver Corporation since May 1968. The plans were to sell the same equipment and expand through possible acquisitions. After a few years in command, Wormley was replaced by Ron Leafblad, a former marketing manager for the WCE.

Many more changes took place over the next few years and resulted in the complete disappearance of the name Oliver and the color Meadow Green in 1976. The WCE took on a new color in 1975 to match the color of Euclid, a company that White Motors had acquired. White Construction Equipment was a division that was phased out during White's downsizing moves when the Mighty-Tow line was discontinued.

Throughout production of the Oliver Industrial machines, Power-Paks, formerly referred to as power units, were sold for a variety of purposes. Units were sold to companies like Tampo and Ferguson to make road rollers. Henke Manufacturing of Janesville, Iowa, was sent the power units of the 70, 77, and 88 to be fit into its road maintainers and shipped back to Oliver to be sold through the company.

Lull Engineering of St. Paul, Minnesota, was one of the many companies that took advantage of the reliable Oliver power units. The company had a good working relationship with Oliver and supplied loaders for many of the Oliver tractors starting with the Industrial 35. It later used Oliver power units, which would be fit into a self-propelled high-lift loader unit.

The Hancock Manufacturing Company was added to the list of companies that used Oliver Power-Paks. This Lubbock, Texas-based company built units such as the 282 elevating scraper with the Oliver power unit in the mid- to late 1960s. It was available with either the GM 4:53 engine or the 310 Oliver/Waukesha engine. Both variations used the instrument panel, firewall, clutch pedal, steering assembly, and Hydra-Power drive directly from the Oliver models. The majority of the Hancock 282 scrapers built used the GM engine, which was much more powerful than the 310 engine. This was the first self-propelled elevating scraper built by Hancock; previous models used a four-wheel tractor-drawn unit.

While Oliver was primarily an agricultural company, it was also a strong presence for many industrial applications. Industrial dealers were scattered throughout the country, but the strongest presence was in Pennsylvania, New York, California, and Washington, where many still survive today.

CHAPTER 4

THE CRAWLER LINE

The Oliver Farm Equipment Company had a well-rounded lineup of equipment available for its customers by the 1940s, but it was missing one important tool: a crawler-type machine.

The leaders in the crawler industry at the time were Caterpillar and the Cleveland Tractor Company (Cletrac). With the post-war boom on the horizon, Oliver officials wanted to expand the line to meet the needs of both the industrial market and the agricultural customer. The Cletrac organization was a family owned and operated business that was looking for a buyer.

Rollin H. White started Cletrac in Cleveland, Ohio, in 1916. Rollin came from a very industrious family. His father, Thomas White, was the founder of the White Sewing Machine Company, which manufactured roller skates, bicycles, and kerosene lamps in addition to its namesake product. Rollin graduated from Cornell University and studied automotive development in Europe. Upon returning to Cleveland, his father gave him some space in his factory for experimental work. Rollin invented a steam car and sold the first one to the public in 1900.

Within the next few years, Rollin's brothers, Walter and Windsor, joined him in the manufacture of steam cars. The automotive division had outgrown its space in the sewing machine factory and moved out on its own in 1904. The White Steamer was quite successful and outsold its competition two-to-one over the next few years, but the gasoline models were starting to take over. White introduced its first gasoline-powered car in 1910.

Rollin began to experiment with a gasoline-powered tractor in 1911. He worked with his design over the next few years and received several patents for his innovations. In 1915, Rollin severed ties with his brothers and the automobile company. Walter and Windsor incorporated their automotive business and became the White Motor Company.

The Model H was just the second model to be built by the Cleveland Tractor Company, but over twelve thousand machines were produced from 1917–19. Although it was expensive, selling for $1,385 to $1,585, it was a reliable machine that set the company up for success.
Author's collection

Prior to Oliver's purchase of Cletrac, many models were painted Buckskin Brown. The Cletrac 55 was introduced in 1932 along with the Model 25. It was the last crawler model built by Cletrac with the Wisconsin engine. *Jeff Hackett*

Rollin turned his full attention to the tractor, incorporating the Cleveland Motor Plow Company in 1916. He applied for his first patent for the crawler tractor that same year and laid the foundation for a very successful company devoted solely to the manufacture of track machines. Many of Rollin's innovations are still used by the modern crawler companies.

By the time World War II rolled around, the company was under the leadership of Rollin's son, W. King White, while Rollin held the position of chairman of the board. King's health was failing, however, and there was no one else in the family prepared or interested in taking over the company.

Cletrac was in a good financial position with net sales of over $23 million in 1943, and King White made it known that he was interested in selling the company. Oliver took notice immediately and acquired the Cleveland Tractor Company on October 31, 1944.

With the acquisition of the crawler business, the Oliver Farm Equipment Company became known as

The Cletrac 30B was built in 1929 and 1930. About 1,450 of these were built during that time and before it was replaced by the 40-30. Weighing in around 7,300 pounds, it was rated at around 38 drawbar horsepower. *Super T*

The Cletrac lineup in the 1930s provided the crawler user a variety of models from the 20-horsepower unit to the mammoth 100-horsepower model. *Author's collection*

The Alfta-Track System was built in Sweden for use in forestry operations. The company, Osa, based in Alfta, Sweden, began importing Cletrac and Oliver crawlers starting in 1942 and converted them to the Alfta track system. Seven hundred units of these units were exported and converted to the track system. A much smaller number of track systems were imported into the United States and converted here. Although initially introduced by Oliver as the Alfta Track, they were later referred to as the Flex-Air-Track tractor. *Author's collection*

the Oliver Corporation. The Oliver logo was changed from the four-company shield to one referred to as the "split shield." Cletrac's 80-acre site in Cleveland, Ohio, was added to Oliver's assets along with thirteen hundred factory and office employees.

At the time of the purchase, Cletrac manufactured five different models of crawlers ranging from 20 to 95 horsepower. This addition to the Oliver line left the firm with a well-rounded selection of power equipment.

When Oliver acquired the Cleveland Tractor Company in 1944, the HG was one of the most popular models in the lineup. Later crawlers took on the OC designation, and the HG became the OC-3. Powered by a Hercules IXB3 engine, the little machine put out 22 drawbar horsepower. *Super T*

Three women seated on the bench of the FD shows the size of the machine. A fleet of these units was sold to the Government of India, and the Cleveland lot was lined with them in preparation for export. *Author's collection*

Cletrac was well known in the industrial field and easily recognized by the same bright orange paint used on every model. Oliver was slow to make any changes because of Cletrac's notoriety.

The basic models at the time of Oliver's purchase included the AG-6, AD, BD, BGS, DG, DD, FD, FDLC, and the smallest ever-popular model in the lineup, the HG; all were powered by Hercules engines.

The AG-6, originally named simply "AG," had been in production since 1936. Materials were scarce during World War II, so Cletrac turned to Continental to provide engines for the AG. With the addition of the six-cylinder Continental gas engine, the model designation was changed to AG-6. This model was manufactured until 1957 with approximately 4,800 units built.

The AD was powered by a diesel variation of the Hercules engine. Production of this crawler continued until 1959 with approximately 4,400 units built.

In the B series, the BGS was produced until 1955, while the BD was produced until 1956. The BG used a JXC Hercules engine and became known as the BGS when the engine was upgraded to a JXD in 1945.

The D series consisted of the DG with a gas engine and the DD, which also used a Hercules powerhouse. The DD was a favorite among farmers for heavy tillage work. It was capable of pulling a six- to seven-bottom plow in most conditions. It also served as a logger model and was popular for road construction and snow removal. The gasoline engine made for easy starting in cold weather, but it was not very economical in terms of fuel usage.

The FD and FDLC were the heavy hitters in the 1944 lineup. The FD was powered with a DHXB Hercules engine. It produced 107 maximum belt horsepower at the Nebraska test and weighed in at 30,000 pounds.

Cletrac turned to the Cummins Engine Company to power its FD in order to give the F series more power. With the addition of the HBISD Cummins engine, this model became known as the FDLC. With its supercharged engine, this model produced 150 horsepower at 1,800 rpms. Only 136 of these mammoth machines were built before the Oliver Corporation discontinued the model.

The HG was a very popular model with nearly 30,000 built from 1939 to 1951. This 3,500-pound

The 64-inch gauge HG was by far the most popular of the HG models. With the wide weight distribution, the machine was well suited for hillside work and quite stable. *Jeff Hackett*

Often models were customized to fit special applications. Such was the case with this OC-6, which is used for spraying onions on a vegetable farm. *Super T*

The OC-4 B series was available with a gas or diesel Hercules engine. The gas engine produced 24 horsepower. The OC-4 diesel was the smallest diesel-powered crawler on the market and cranked out 26 horsepower. *Jeff Hackett*

The OC-4 used three different grille styles throughout production. The first model used the same grille as the Fleetline series of tractors. This is the second style used to give the OC-4 a more industrial look. *Jeff Hackett*

model used a four-cylinder flathead Hercules gas engine that delivered 20 maximum horsepower by the belt. It was offered in four different gauge widths ranging from 31 to 68 inches. This machine was originally Cletrac orange, but Oliver initiated a color standardization program in 1946. Since the HG was used primarily for the agricultural market, the color was changed to green to match the rest of the equipment line. The rest of the Cletrac crawlers in the lineup remained orange throughout their production.

Oliver did little to change anything for several years due to Cletrac's excellent reputation. The models were becoming outdated, however, which resulted in a loss of sales. Caterpillar, A-C, and IH took over a large portion of the market share. Since Oliver/Cletrac didn't have the sales volume of its competitors, it didn't have the purchasing power, which made the Oliver/Cletrac models more expensive.

Oliver discontinued the use of the Cletrac name in 1950. All crawlers would now be known as Oliver crawlers, and a new series was born that sported the OC (Oliver Crawler) designation. The HG crawler was renamed the OC-3 and followed in HG's footsteps by taking on the Oliver green colors through a portion of the early 1951 production. At that time, Oliver decided to change all of its industrial colors to Highway Yellow. The larger models were also changed to this color. The OC-3 was built from 1951 to 1957 with approximately 20,000 units produced.

Oliver introduced the OC-6 crawler, a model designed specifically for farm use, in 1953. This model was styled after the Oliver 77 wheel tractor with the same sheet metal and Oliver/Waukesha engine. The company even sent out a service bulletin stating that if a loader or blade were mounted on the OC-6, the warranty would be void.

The next machine in the lineup was the model OC-4. This first variation was known as the A series and used the same engine as the OC-3. It was styled with the Fleetline sheet metal to match the look of the OC-6. The OC-4 was available in five different track widths to give Oliver a wide variation of models produced from one basic machine.

Although the appearance of the OC-4 was sleek and matched the current line of tractors, the sheet metal was not durable enough to stand up to the rigorous use of industrial applications. A heavy grille assembly was designed and incorporated on the next series of the OC-4 crawler. This design also marked the introduction of the OC-46, an OC-4 with a mounted loader that became quite popular with small contractors. It was a durable and well-balanced machine that wasn't simply converted to fit the purpose, but rather was designed from the ground up to meet the needs of the small contractor.

Oliver introduced its spot-turn steering system in 1959. Its clutch-type steering was simple, efficient, and safe. The clutch and brake were simultaneously activated with just the pull of a lever. The machine

The Industrial Cab Company of Salem, Massachusetts built this cab, which is mounted on an OC-4 sidewalk plow. These cabs were not available from the Oliver Corporation, but the dealer could order them directly from Industrial and mount them at the dealership. In 1965, Industrial Cab Company moved to Flora, Indiana. *Author's collection*

Cabs were an option offered by the company but were not actually built by Oliver. Several aftermarket companies built cabs for a wide variety of companies. The Dole Cab Company of Willimantic, Connecticut, provided the cab on this machine. *Jeff Hackett*

instantly stopped when both levers were pulled at the same time. The machine had a turning radius of 5 feet and 4 inches with this unique system.

The next model to come out of the OC line was the OC-18. This was the largest machine in the OC series, weighing in at 32,500 pounds. The OC-18 was powered by a Hercules DFXE six-cylinder engine and was rated at 133 horsepower on the drawbar and 161 on the flywheel. The OC-18 was larger than the Cat D-7 but smaller than the D-8.

The power-turn system was introduced on the OC-18. This steering method was the best of both worlds; it incorporated the finest features of clutch and differential steering. With this system, one side could be braked without declutching or taking power away from the opposing side. A slower, more controlled turn could be achieved by slowing down one track while braking.

Air steering was a unique feature of the OC-18. This was originally incorporated as an option on the FDE models. The government of India ordered large quantities of this model for clearing land, but the model's size presented some difficulties. Fasting

The largest model in the OC series was the OC-18, which weighed in at 33,000 pounds. This machine was powered by a six-cylinder Hercules engine and put out 150 horsepower when tested in Nebraska during 1952. Around 800 of these machines were built before production ended in 1960. *Author's collection*

was an integral part of Indian custom, and operators were often too weak to maneuver the heavy machine. Because of this, Oliver incorporated an air-assist steering unit on the FDE for easier handling. The steering was carried over to the OC-18 as standard equipment.

The OC-12 was the next model to hit the scene and was designed as a replacement for the Cletrac B series. The gas model appeared in 1954, and the diesel variation appeared in 1955. Both models used Hercules engines and were equipped with Cletrac's famous controlled differential steering unit. This system, which put power to both tracks at the same time, was replaced in 1959 with the spot-turn system that allowed for more controlled power with shorter-radius turns.

When a loader was mounted on the OC-4, it was designated as an OC-46. This also held true for the OC-12. The OC-12 was transformed into the OC-126 with the 1 1/2-yard Ware loader mounted on the 60-inch gauge crawler.

The OC-15 and OC-156 were introduced in 1956 as replacements for the DD and DG. The engine of choice, the Hercules DRXC, delivered 94 horsepower on the drawbar. These machines weighed in at 19,000 pounds and Oliver boasted that they had the greatest horsepower to weight ratio of any crawler built in that class.

These models were equipped with standard air-assisted steering. Oliver added its spot-turn clutch steering system in 1959 to help with the shorter pivots during loader operation. The Ware bucket mounted on the OC-156 had a 2 1/4-yard capacity with a maximum lift of 12,000 pounds. A 3,500-pound counterweight was mounted on the rear to keep the rear of the machine on the ground. The OC-156 was the largest crawler loader produced by Oliver. Even though this loader was superior to others on the market at the time, sales were low, and less than 100 units were built from 1956 to 1961.

The last crawlers to be added to the OC series were the OC-9 and OC-96 in 1959. Although they were designed to replace the A series, the OC-9 and OC-96 were completely new machines. Because they were designed specifically for industrial use, they were never tested in Nebraska like previous models.

The OC-96 was powered by a Hercules four-cylinder engine, rated at 57 flywheel horsepower, and weighed in at 14,150 pounds. The Trans-O-Matic transmission incorporated a modern torque converter for maximum efficiency, which made it very popular with contractors. *Author's collection*

The bare OC-9 machine weighed in at 9,675 pounds. When it was equipped with a dozer blade, it jumped to around 13,000 pounds. The OC-96 weighed 14,750 pounds and was equipped with a 1,000-pound counterweight. Oliver also introduced its new Trans-O-Matic power shift transmission with the OC-9 and OC-96. This feature, combined with power steering and instant reverse features, gave these machines superior maneuverability.

The transmission used in the OC-9 and OC-96 was built by the Transmission and Gear Company of Detroit, Michigan, and featured two speed ranges in forward and reverse. The Hercules four-cylinder engine was coupled to a Clark torque converter and eliminated the need for a master clutch.

When White Motors bought the Oliver Corporation in 1960, White had different ideas about the operation of the crawler division. The crawler line was not brought into the new White family but was separated and named the Cletrac Corporation. It operated separately to build crawlers in the Cleveland plant that would be sold exclusively through the Oliver division of White Motors. Oliver rented the plant from the Cletrac Corporation and reserved the option to buy the plant, which expired in 1962. Cletrac, on the other hand, made Oliver agree to match a certain dollar figure in sales per year or the deal would be off.

On November 1, 1961, White decided to purchase only certain assets of the former Cletrac line for $5.5 million. White did not acquire the Cletrac name, the Cleveland plant, or the crawlers themselves, but did acquire the models OC-4 and OC-46, along with the OC-9 and OC-96. These were the only models left

The FCG with a blade, operated by the Wm. Lathers Company from Madison, Wisconsin, was used in Reedsburg, Wisconsin, to level dirt and rock. This was most likely a fill job for a new roadway. *Author's collection*

in production at that time.

White Motors was a truck company, and although it was looking for an agricultural line to add to its portfolio, it wasn't dedicated to crawlers. Once White Motor had acquired what it wanted of the crawler line, it transferred production to Charles City. The OC-4 and OC-46 took on a new look with a heavy-cast, industrial front. This model became known as part of the B series, along with the newly improved OC-9 and OC-96.

The crawler engineering department in Cleveland had new models that were going through experimental testing. These machines, had they been produced, would have been modern machines that would have provided serious competition for the others. Little money was allotted for new tooling once production was moved to Charles City, and even though the new models would have changed the outlook of the crawler division, White opted to make minor improvements to the current line.

Oliver was focused on big tractors and felt that the four-wheel-drive model would soon make the crawler line obsolete. Oliver deserted the Cleveland plant; its last crawler was built in 1965. Production of Cletrac and Oliver crawlers came to an end after forty-nine years. Still, the four-wheel-drive tractors do play an important role on the industrial scene today, and the crawler is in no way a part of the past either.

OLIVER
550

CHAPTER 5

OLIVER IMPLEMENTS

The lengthy lineup of Oliver equipment lasted well over a century. The changes made in the last one hundred years of Oliver revolutionized farming. Plowing changed from a chore that was labor intensive to one that consisted of simply turning a key and sitting behind a steering wheel. The rule of thumb for plowing one hundred years ago with a horse was one acre per horse per day; therefore, a three-horse team could plow three acres a day. The team also required watering, resting, and feeding, not to mention the wear and tear on the teamster. With a tractor, one man could go to the field for a day and plow over six acres per hour. At the end of the day, all he had to do was add fuel and go home for the evening. The tractor didn't need to be groomed or bedded down, and the driver wasn't exhausted.

Although James Oliver first entered the foundry business in 1855, he didn't add his chilling process to the plow until 1857. Early plows were built as walking plows only. Most of these consisted of a heavy oak beam. Later plow frames made of steel were added to the options.

When Oliver began casting plow moldboards, they were smooth on both sides. In 1874, Oliver was granted a patent on an improvement in chilling in which grooves were cast onto the backside of the moldboard. The design added strength to the moldboard and prevented it from twisting and warping during the cooling process. That same year Oliver was also granted a patent that covered the process of casting beveled bolt holes in the moldboard.

The year 1874 was very busy for the Oliver company; it purchased new land for expansion, discontinued the manufacture of wagon skeins and Singer Sewing machine castings, and decided to devote its full attention to tillage tools. This vast expansion program scared many of the investors who thought Oliver had bit off more than it could chew. Nearly

The 3241 plow was the perfect companion for the Oliver 550 utility tractor. Many of these models are still in use today and are the ideal size for plowing small patches. *Jeff Hackett*

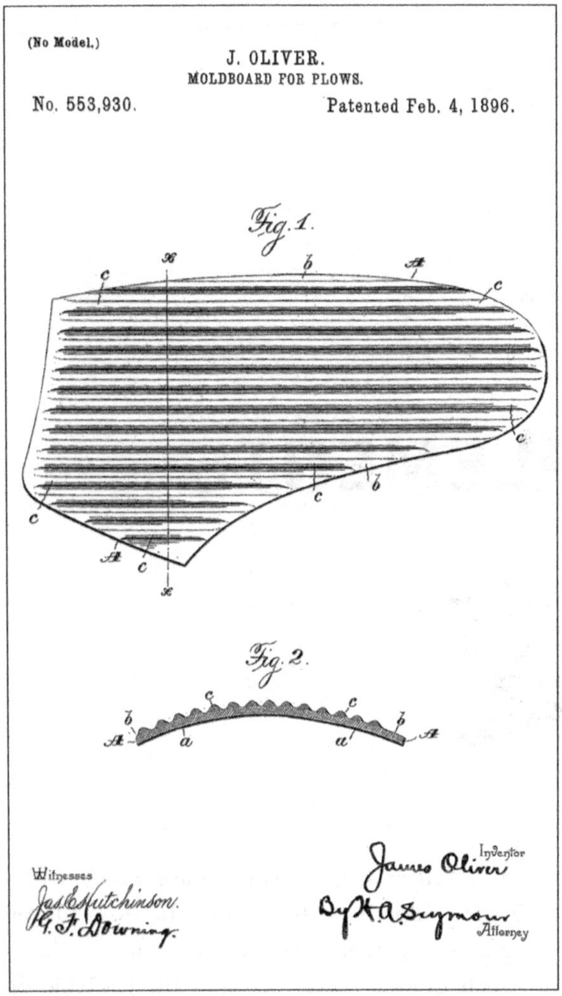

In 1896, James Oliver patented his moldboard with the ribbed back side that eliminated warping during the chilling process, keeping a uniformed shape.
Author's collection

one-third of the capital stock of the company changed hands that year due to the investors' fears.

The most popular walking plow in Oliver's history was introduced to the farmer in 1874. Until that year, the No. 40 had never been tested in public nor advertised, although it had been produced for a few years. The No. 40 was a two- or three-horse plow that was available as a wood- or steel-beam model. As people drive on different sides of the road in different parts of the world, they also require different plows; therefore, Oliver offered both a right- and left-handed plow. The 40 was manufactured for nearly eighty years with minimal improvements.

In 1876, Oliver introduced its first sulky plow: the Casaday. Although William L. Casaday patented it, the plow was built and marketed by Oliver. By 1886, Casaday was no longer associated with Oliver and took his plow with him. Oliver began to manufacture its own sulky plow, which was known as the "New Oliver."

Oliver began the design of a walking-disk plow in 1900. By 1903, it had added a disc gang plow to its line. This plow allowed corners that hadn't been previously accessible to be plowed. Other products were added to the line, such as walking planters, riding planters, and riding cultivators. A two-way riding plow was added to the list of 1908 accomplishments.

With the advancement in power farming on the horizon of American agriculture, the gas engine was cutting edge. Oliver could see that the trend was turning, and plows would have to be designed for agriculture of the future. In 1911, Oliver was granted a patent for an engine gang plow, which became the focus for that year. To show the superiority of its plows, Oliver put together a demonstration on its farm in South Bend. Using three Rumely Oil Pull tractors, a plowing world record was established. Ten-section, five-bottom engine plows were hooked together and turned a 60-foot swath. Prior to this, the record had been 21 feet. This showing set the bar for the competition in the plow market. The record didn't hold long, however; later that year, three IH Moguls pulled fifty-five bottoms and turned over an acre in less than four minutes.

During the 1910s, smaller tractors were being built to replace the large, cumbersome, prairie-type tractors. Oliver reacted by introducing its No. 62 hand-lift tractor plow. It was built with two or three bottoms. Most IH tractors built at that time pulled that plow model.

The following year Oliver designed a plow, known as the No. 7 gang plow, to be used exclusively with the Fordson tractor. J.D. Oliver had a good relationship with Henry Ford in the earlier days and designed an entire line of equipment to be used with the Fordson and sold through Ford.

Although the plow was the foundation for Oliver's company, it also offered an entire line of tillage tools. Prior to the merger, it had used other manufacturers

Oliver's foundation was built on the successful line of plows. The 4240 plow was offered with the rooster-comb or hydraulic lift. This particular plow has the long axles, indicating it is an early model. *Jeff Hackett*

to supply the equipment that it couldn't build itself. The Thomas Hay Tools company supplied rakes, tedders, swathers, and hay loaders during the 1920s. The Superior Drill company supplied the horse-drawn drills and seeders, along with Black Hawk manure spreaders and planters. Oliver sold Massey-Harris mowers, grain binders, hay rakes, and tedders. The Oliver Chilled Plow Works also sold Swayne-Robinson hay balers and ensilage cutters, as well as Cope stackers and rakes.

Wagon gears were built by the South Bend Wagon company. These wagons used an oak running gear, hickory axles, and a box that was made of Mississippi cottonwood. Oliver also sold a stylish buggy built by the Sechler Carriage Company. While not many examples of this buggy exist, there is a restored model on display at the Heartland Museum in Clarion, Iowa.

Different Oliver branches offered wagons from different manufacturers. Oliver also sold products for Winona, a wagon builder in Winona, Minnesota, and the Peru Truck company in Peru, Illinois.

When the Oliver Chilled Plow Works merged with the other companies in 1929, it had many of the previously mentioned products being built in its own line. The company didn't need a jobber to fill in the gaps in production.

In 1939, Oliver introduced the Raydex bottom, which was a hardened, throw-away share. Because of its low cost, it was easier to replace than sharpen. A new share always had the factory cutting edge, good penetration, and a perfect fit. The Raydex points were as sharp as a razor blade, and Oliver gave away Raydex razor blades with the purchase of a box of Raydex points.

Oliver introduced the Spee-Dex bottom in the late 1960s. This new sculptured bottom lightened draft by 10 percent and increased productivity by as much as 20 percent. It was possible to plow up to 7 miles per hour, depending on field conditions. The farmer

The Oliver No. 23 Sulky Plow was originally designed for use on hillsides where it was necessary to plow back and forth in the same furrow. Later, it gained popularity on level land, as it left no dead furrow or back furrows. *Author's collection*

The Superior drill came into the Oliver line when the companies merged in 1929. This drill was one of the most successful products in the American Seeding lineup. *Super T*

could use a bigger bottom to turn over more ground per day or use the same bottom and plow faster.

The Oliver plow, noted for its light draft and good clearance, was very popular in the farm equipment market. It was built in more configurations than can be mentioned—orchard, corn bore, road, paving, and hillside plows, to name a few.

Oliver continued to build moldboard plows until soil conservation became a factor. Moldboard plowing was replaced by no-till methods, but its success is evident by the hundreds of models of plows built in the 120-year production span.

When the merger of Oliver, Hart-Parr, Nichols & Shepard, and American Seeding took place in 1929, some of the outside product lines were dropped and replaced with ones within the new company. American Seeding brought its grain drills, seeders, and manure spreaders into the family.

The Oliver Superior Drill was built in a wide variety of sizes, from this large model down to a 3-foot-wide, horse-drawn walking model. These drills were very successful and were a common sight on every farm. *Jeff Hackett*

The 100 baler used a smooth platform on the deck to prevent the loss of leaves. The elevator fingers were mounted above the hay to help feed the cross conveyor. *Jeff Hackett*

The Oliver 100 was an automatic wire tie baler. A man no longer had to ride the baler to tie the knots. The chamber on the 100 was 16x18 inches and operated at 38 strokes per minute. *Jeff Hackett*

The planter evolved from a one-row walking, horse-drawn machine to a sixteen-row air planter. When Oliver incorporated its pipe-mount system into the new tractors in 1930, it also introduced implements that could be used on the tractor rather than behind it. Single planter units could quickly be attached to the pipes that ran through the frame of the tractor. Two units could be mounted on the smaller tractors to create a two-row planter. Larger tractors used a four-unit system that planted four rows. Fertilizer boxes that dropped dry fertilizer into the row while planting were also attached to these units.

Oliver manufactured a few three-point planter units for the smaller tractors after the advent of the three-point hitch system. The 310 was a two-row planter designed for use with the 550 tractor. It was adjustable from 34- to 44-inch row spacing. The 310 evolved into the 312 in 1959. An implement that starts with the number "three" is typically meant for use with a three-point hitch system.

The Oliver Farm Equipment Company began selling Superior planters and drills after the merger. American Seeding had purchased Superior Drill prior to the big merger, but the Superior name was very well known. All of Oliver's drills and planters carried the Superior and Oliver names until early 1950 when the Superior name was dropped. Oliver drills were manufactured into the 1970s, the largest of those was the Model 96.

The early manure spreaders Oliver sold were originally sold by American Seeding and marketed as the Oliver Superior manure spreader. The first spreader introduced by Oliver after the merger was the Model 75, which was built in Springfield, Ohio. It was

The majority of Oliver's wagons were built by Electric, which was located in Quincy, Illinois. Although this barge box was home-built, the running gear is Electric and decaled as Oliver. *Jeff Hackett*

Electric Wheel produced the wagons for Oliver and offered a complete line of boxes for its running gears, such as the gravity box, flare box, and barge wagon with removable sides. *Jeff Hackett*

When hydraulics were introduced on tractors, wagon manufacturers took advantage of it and equipped their dump running gears with a lift for self-dumping. *Jeff Hackett*

a wooden box spreader with a 75-bushel capacity, thus the designation of Model 75. This model went through various changes throughout its production; one of the most notable of which was the addition of the inverted arch. This feature took the supporting arch that was above the reel and flipped it upside down. When the arch was located above the reel, it interfered with the reel and manure hung up on it during loading. The arch no longer caused interference when it was flipped upside down and still gave the spreader the support it needed.

Oliver introduced its first spreader designed to be used with rubber tires in 1939. This model was the No. 7 and used a steel ribbed box with a 77-bushel capacity. A smaller variation was the No. 11, a smooth-sided steel spreader with only a 50-bushel capacity.

Previous models were all ground-driven spreaders. Oliver introduced its first PTO-driven spreader in 1954. The Model 100 was the first of these models and had a 100-bushel capacity.

The largest Oliver spreader was the 580 model. This PTO-driven, tandem axle unit had a capacity of 240 bushels. It used a single Saw-Padl beater that could unload the full 240 bushels in as little as 1 1/2 minutes, and due to the width of the machine, it used a twin conveyor system. By the mid-1970s, Oliver wasn't really interested in being in the manure-handling business anymore and phased out the spreader line, which had been built in South Bend, Indiana, at that point.

The Oliver baler line was the result of the Tallman family of Shelbyville, Illinois. Horace Tallman was the owner of the Ann Arbor Machine Company. He had been in the hay business most of his life, and his mission was to design the best baler possible. He contributed many innovations to the baler during his life. He conceived the idea of a pickup baler but passed away in 1929 before the machine was completed. His sons, Leslie and Lloyd, carried on their father's dreams and had the new hay and straw combine on the market the next year.

Oliver's acquisition of the Ann Arbor product line in 1943 added a very popular baling machine to Oliver's portfolio. Oliver went on to develop a full line of balers that were manufactured well into the 1970s. Mounted engines, either Continental or B-4 Wisconsin, powered the early pickup balers. It wasn't until later in the 1950s that Oliver's balers became PTO driven.

Oliver had an entire line of hay equipment that consisted of rakes, mowers, and balers, all designed and manufactured in the Shelbyville plant. Both pull-type and mounted sickle mowers were offered. Oliver introduced the Model 437 pull-type mower/conditioner in 1967. This was a dual-purpose mower

The Raydex plow share was first introduced in 1939. Similar to the chilling process, it provided a hard durable surface. By the late 1950s, Raydex had evolved into "Super" Raydex. *Jeff Hackett*

Herman Algelt, an engineer at the Oliver plow plant in South Bend, invented the pipe-mount method of attaching an implement. This system was devised by Algelt before the merger with Hart-Parr and was so innovative that the company incorporated it onto all row crop models for the next 30 years. *Jeff Hackett*

that cut and crimped the hay as it came out the rear of the machine. The 438, capable of cutting and crimping the hay as well as laying it in a windrow upon exit, was introduced the following year. When the 129 mower made the scene, it was equipped with either steel or rubber rollers.

Nichols & Shepard, based in Battle Creek, Michigan, was the manufacturer of the Red River Special line of threshing machines. When the big merger took place in 1929, the Oliver name was added to the already-popular machine. The Nichols & Shepard Company had been manufacturing harvesters since 1848, so it didn't lack experience. This made the company the perfect choice to act as the harvester division of the newly formed Oliver Farm Equipment Company. Nichols & Shepard's threshing machines were among the best in the industry and were manufactured until 1950, by which time the pull-type combines had taken over in the harvesting process.

With the introduction of the three-point hitch, Oliver also introduced several different implements unique for that application. *Super T*

The first pull-type combines were built in 1935. Only 161 of them were built in three different variations. Most of these were used for testing, and 1,070 were built in 1936. The numbers continued to grow from there.

Oliver began production of the Model 15 pull-type combine in 1946. This machine incorporated a new unloading system known as an auger. Combines had to use an overhead tank with a gravity flow to unload before the auger was introduced. The unloading auger mounted on the 15 allowed the operator to pull up to a wagon and dump the extra-large, 24-bushel hopper.

It was 1950 before Oliver came out with its first self-propelled combine, the Model 33. It was followed later that decade by the 35 and the 40, which used some of the same components that were used in the Oliver tractors. The Model 35 was the basic grain combine, but the Model 40 could do both small grains and corn.

In 1959, Oliver introduced the Model 25 as its new machine. There were 101 units built, half painted red as a Ford machine. After all the bugs were worked out, the 1960 production of these two machines was 1,200 units. Production of the Model 25 was moved to Brantford, Ontario, after White Motors acquired the Cockshutt plant. It continued to produce the combine for Ford, known as the Ford 611; however, Ford combines that came out of Brantford were now painted blue.

The Oliver combine engineers began planning for a variable-speed drive transmission in 1954 for the Model 25 combine. Talks began with the Vickers Company, and two machines were equipped with the Vickers system and sent out for testing.

The Superior line consisted of spreaders as well as grain drills. The Number 11 had a 50-bushel capacity when rounded up. It could also be ordered as a four-wheeled horse-drawn unit. *Super T*

At one time, the operation of a baler required an entire crew of men. Ann Arbor revolutionized that task and eventually eliminated everyone but the driver of the tractors. *Jeff Hackett*

Ann Arbor was one of the most recognized names in the baler industry. Not only did it build the balers for the individual farmer, it also provided commercial balers. In 1930, it became the first company to manufacture a pickup baler. *Jeff Hackett*

Oliver began working with the Sundstrand Corporation in 1960 and mounted the system on a third experimental combine. Sundstrand had introduced its experimental hydrostat transmission on the John Deere 45. In order to make production feasible, Sundstrand needed to be able to provide these transmissions to more than one combine manufacturer.

Oliver was ready to go into production with its first hydrostat combine in 1961. Oliver's parent company, White Motors, however, had already decided it was going to close the Battle Creek plant. White had its sights on the Cockshutt combine plant in Brantford, Ontario. In an underhanded corporate takeover in 1962, White Motors became the parent company of Cockshutt, a well-established farm-equipment company in Canada, and replaced the Battle Creek plant.

After combine production was transferred to Brantford in 1962, Oliver's hydrostat model was scrapped.

Even though the Ann Arbor baler began its life in Ann Arbor, Michigan, it called Oliver's Shelbyville, Illinois, plant "home." *Jeff Hackett*

With the one-man automatic-hydraulic threader, only one man was required to feed and knot the wire on the baler. The clinch head bale tie was the answer to knotting the wire. This was introduced in the late 1930s on the Ann Arbor baler, Oliver's predecessor. *Jeff Hackett*

The 720 baler was built from 1967 to 1972. It was the replacement for the 620 and offered more capacity for the larger hay operations. In addition to being built in Meadow Green colors, the same baler was also built in Minneapolis-Moline and Cockshutt colors. *Super T*

The twin-gear setup on the baler provided more power behind the plunger and made the machine more balanced.
Jeff Hackett

Oliver came out with a hydrostat combine in the late 1960s, and by that time, other manufacturers were already using that feature.

Cockshutt's experienced combine engineers took over design of the Oliver combines once production was transferred to Canada. Cockshutt was already tooled up to build the 430/431 combines when Oliver stepped in. Those same machines were built for Oliver but were painted green. Models built as Cockshutts that were for sale in Canada were painted red and sold as Cockshutt combines. Since White Motor also owned Minneapolis-Moline, combines were also painted yellow and tagged as Minneapolis-Moline models.

Many implements were produced by Oliver and sold only in certain areas. Oliver acquired Farquhar, which offered a variety of orchard sprayers, many of which ended up in Florida and California. Potato planters were also sold as Farquhar-built units but today are rarely seen anywhere other than the East Coast. Both of these lines were backed by experience and a good reputation.

Although companies didn't want the consumer to know that its products were built elsewhere, Oliver built some equipment for the competition. The company not only built plows for Minneapolis-Moline, its sister company, but also for Ford.

At the time of its 1929 merger with Oliver, Nichols & Shepard was experimenting with a two-row, pull-type corn picker/husker that would take the place of a dozen hands helping farmers to evolve with the transforming agricultural industry. This machine

The 100 Automatic Baler, built in Shelbyville, Illinois, was sold from 1953 to 1958. This machine was available with either a Wisconsin or Continental engine. In 1955, list price on the model equipped with the Continental and electric start was $2,419. *Jeff Hackett*

The No. 2 Rake was the first Oliver rake to be designed for use with a tractor. It was a high-speed model equipped with a windshield to keep the hay from blowing and twisting when being raked. The No. 2 sold new for around $300 in 1950. *Jeff Hackett*

Many implements were made available for the Oliver utility line. When Oliver had a void in the line, it would outsource it to another company and have it branded as an Oliver. Such is the case with the Oliver 960 mower, built by Danuser. *Super T*

Oliver's hay division in Shelbyville, Illinois, manufactured a successful line of hay rakes to be used in conjunction with their mowers and balers. The No. 2 was produced through the late 1950s and was replaced by the oblique angled No. 107 rake. *Jeff Hackett*

The Red River Specials were a very successful and popular line of threshers introduced by Nichols & Shepard and built in Battle Creek, Michigan. *Author's collection*

The Red River Special was introduced by Nichols & Shepard and was a trouble-free unit that sold in large numbers. Oliver offered threshing machines until the late 1940s when they were phased out by the pull-type machines. *Jeff Hackett*

was revised in 1933 to become the Model 2 Corn Master. It was built until 1953, going through many updates along the way. This was the most abundantly produced picker in the Oliver line with over twenty-four thousand machines built. When this unit was phased out, it was replaced by the Model 3, which was also a two-row, pull-type picker.

The one-row Corn Master No. 1 picker was built in low quantities from 1939 to 1941. Production ended at the beginning of World War II and did not resume. Oliver introduced the new and improved Model 5 in 1947. This model was built in Battle Creek until Oliver closed the plant in 1962. Production was then transferred to the Shelbyville, Illinois, plant. The Model 5 was built until 1969 in both the Oliver and Cockshutt colors. The Model 4 was a tractor-mounted, two-row picker. It was introduced in 1950 for use with the Fleetline series of tractors. This model was produced until the Battle Creek plant closed.

In 1955, the engineers in Michigan designed a single-row picker known as the Model 6 that semi-mounted on a Super 55. This was not a popular model and few have been saved.

The Model 73 was the last picker to be designed at the Battle Creek plant. This two-row machine, painted as both a Cockshutt and Oliver model, picked,

Combines were an important part of the Oliver farm equipment line. Prior to 1962, the machines were built in Battle Creek, Michigan, but many accessories, such as heads, were built at Oliver's hay division in Shelbyville, IL *Jeff Hackett*

The Model 8 was built from 1949 to 1952 and powered by a Wisconsin engine. It was then replaced by the Model 100, which had twice the capacity of the Model 8. *Author's collection*

The first pull-type combines required an additional operator to run the machine. Even with the two-man operation, this was light years beyond the early threshing bee days. *Author's collection*

husked, and had the capability of use as a sheller. The first model was built in very limited quantities for testing in 1957. Production was dramatically increased in 1959. After the Battle Creek plant closed, Shelbyville's plant took over production until 1965. The Minneapolis-Moline version was also built there but used the number designation of H320 or S320, depending on if it was a husking or shelling machine.

The remainder of corn picker production was done in Shelbyville and consisted of the Models 74 and 83. All picker production ended in 1972. The combine had taken over the chore of picking, husking, and shelling.

Farm equipment companies were struggling during the 1970s, and meanwhile, White Motors, a truck manufacturer, was holding millions of dollars of bad debt due to repossessed trucks. White did everything it could to consolidate and cut costs. Plants were closed, and production was transferred to other plants. Equipment that was not high in volume sales was sourced out. Oliver still offered a full line of equipment, but some of it wasn't its own. The farm equipment business had changed drastically during the twentieth century.

OLIVER
880
OLIVER

CHAPTER 6

SPECIAL TRACTORS

Oliver introduced its Teamed Power Parades in 1959. This introductory parade was the debut of the 1959 wheel tractor lineup, which included the 550, 660, 770, 880, and the 900 series tractors.

An official Oliver sales note written on January 6, 1959, but dated February described the new color of the 880.

"As you examine the 880 for 1959, you will notice a definite change in appearance. The color is different. The new pastel shade of green gives you the advantage of having a larger appearing tractor to sell. Keep in mind that this new color, Mist Green, is for the 880 only."

The 770 and 880 tractors were hard to distinguish from a distance. Oliver felt like the pastel color on the 880 would help set it apart from its little brother. The first mist green 880 rolled off the assembly line on January 23, 1959, with serial number 73639. The tractor looked faded and mismatched every piece of equipment built by Oliver. The helical constant mesh gears were introduced at this same time.

On March 23, 1959, Oliver issued a sales note titled, "Change in Agricultural Sales Note No. 63" that stated:

> *"The management of the Oliver Corporation has made two changes which affect remarks made in the February sales note . . . The changes were announced after the bulletin was prepared.*
>
> *"The management has decided that there will be no change in the 880 tractor. It will have the same color scheme as in the past: Meadow Green and Clover White.*
>
> *"Therefore, we ask you to disregard comments we have made in the sales notes and during the recent 'Teamed Power Parades' concerning these particular items. We sincerely hope that you will not be inconvenienced by these changes."*

The most unique, but correct, paint scheme that people mistake for a faded tractor is the Mist Green 880. This tractor is still in the family of the original owner and restored to the exact color. When introduced in 1959, potential buyers did not welcome the color and many were painted back to Meadow Green. *Jeff Hackett*

The color difference can easily be seen when sitting alongside a true Meadow Green Oliver. The Mist Green 880s were built from serial number 73639 to 73891, which makes it the rarest of the 880 models. *Jeff Hackett*

The rarest variation of the Oliver 70 is the Airport model, designated as the 25. These special tractors were used by several airlines such as Transcontinental, Western Airlines, Mid-Continent Airlines, Delta, American, and Pennsylvania Central Airlines. *Jeff Hackett*

The last tractor to be produced with the Mist Green paint scheme was serial number 73891 on January 29. A total of 249 of these models were produced before the paint returned to the Meadow Green color.

Airport 25

The Airport 25 was introduced in 1939. Although it is designed after the 70 Industrial, the earlier serial tags designated it as a 25. In addition, it had some unique characteristics that would be useless in other applications. Models were equipped with an automatic coupler which made for quick hitching up to an aircraft. A snatch block hitch was another feature that assisted in the moving of aircraft. Several companies used these tractors during the early 1940s. The early models were painted Vermillion red, like the Cockshutt 70. The later 25s were painted an industrial orange color.

Author's collection

The 25 was also marketed to the airports for use in mowing, plowing snow, and sweeping ramps and runways. According to the sales literature, the 25 could remove up to 8 inches of snow and clean as much as 20,000 square yards in an hour at a tractor speed of 6 miles per hour.

The Specials

The farm equipment market was starting to experience low sales in the late 1960s. Because tractors were lasting longer, fewer were being sold. Corn prices were low, and those who might have needed a tractor weren't willing to pay the high price. Even John Deere, who had record sales for the past six consecutive years, experienced a drop in sales.

John Deere had to take action; it had just spent $1.5 million on a massive plant expansion. Deere set out to market an economy-model tractor in order to boost sales. The JD 4000 was the answer to tough times. This model had the same power as the 4020 but without the frills and was advertised to "compare favorably with competitive 76- to 80-horsepower tractors." International Harvester's answer to John Deere's 4000 was the Custom 856.

Oliver reacted to this move with a line of its own. The May 1969 price books show the introduction of the "mini-equipped tractor." These models were available with a choice of engines, a standard transmission (six speeds forward and two reverse), Tilt-o-Scope steering, pressed-steel front wheels, and cast-iron rear wheels.

The mini-equipped models were available as the 1550, 1650, 1750, 1850, 1950, 1950-T, 2050, 2150, and 2150 EHD. Oliver took one more step and slashed options and prices even further two months later.

Oliver introduced a model in its price sheets in July 1969 that was known as the "Special." This model

A July 1969 sales note pushes the sale of the Special models during tough economic times.
Author's collection

Sales Notes

LOCATION Charles City
SUBJECT 1750-1850 Specials
NUMBER 433 610
DATE July 1969
FILE 1 of 2

1750 SPECIAL / 1850 SPECIAL

80 Certified pto Horsepower / 92 Certified pto Horsepower

Current economic conditions have caused the equipment industry to focus attention upon a new marketing method—selling to the farm operator who places more emphasis upon proven horsepower capabilities than he does upon extras.

Figure 1

Figure 2

To help you gain a share of this market, Oliver has introduced two new models— the 1750 SPECIAL (Figure 1) and the 1850 SPECIAL (Figure 2). No change in important performance characteristics because both models have retained *same* engines, *same* drive trains, and *same* draft-controlled hydraulics as regular production 1750's and 1850's. Eliminating some extras and standardizing production means that you can now pass extra savings along to your customers.

REMEMBER—You can sell same performance—full drawbar capability, PTO operation, draft-sensing hydraulics—but at lower cost.

The mini equipped Special tractors used a tinted white paint in place of the Clover White. This paint is often mistaken as a paint job gone bad but is an accurate representation of these models.
Jeff Hackett

was only available as a 1750 Row Crop and 1850 Row Crop. These were very basic models that were only equipped with a diesel engine, tilt-only wheel, adjustable front axle, a single headlight mounted on each fender, and pressed-steel wheels. The basic price for the Special 1750 was $7,562, while the regular 1750 diesel model had a base price of $8,254. The basic price for the Special 1850 was $8,260, while the regular 1850 diesel model was based at $8,952.

The Specials are easily recognizable. They used Meadow Green for the primary color and a darkened version of Clover White for the accent color. The word "Special" was decaled on the hood. This model should not be confused with the 550 Specials or the Corn and Soybean Special series of combines.

Although Oliver didn't put this machine on tracks, it would be at least ten years before a track machine was added to the product line. *Jeff Hackett*

This very rare model is the 18-28 equipped with Trackson tracks. Trackson, located in Milwaukee, Wisconsin, often fitted its tracks on various machines or offered aftermarket kits. The most commonly seen Tracksons have Fordson power units due to the low cost of the Fordson. The combination of an 18-28 and Trackson tracks most likely made it an expensive machine. *Jeff Hackett*

The HGR was billed as an all-year-round tractor adaptable to use in snow, mud or marshland. Unfortunately, the technology wasn't around yet to produce rubber of a consistency that was durable enough for this machine. Tracks would stretch and slip off the pulleys on a regular basis. Oliver eventually recalled this model and converted them all back to steel tracks. *Author's collection*

When Oliver introduced the first of the Fleetline series, it had not yet introduced the new sheet metal styling, so the early 88 took on the look of the models 60 and 70. These were only built in 1947 and early 1948. Oliver updated the sheet metal to give it the modern appearance, so it didn't look so much like the old models. *Jeff Hackett*

Hi-crop models were mostly used for cultivation, but special implements were needed due to the height of the machine. These were most popular in vegetable farming while crops were gaining height but still needed cultivation. *Super T*

In 1956, the Kansas Turnpike Authority placed a special order for thirty-two Super 77 tractors to be equipped with DANCO belly mowers. These were to be painted in the Kansas Turnpike colors of blue and yellow while at the factory. *Jeff Hackett*

The Orchard models were configured especially for use in orchards and vineyards. Anything above the hood was lowered to clear the limbs, including the operator's station. Orchard fenders usually took a beating from limbs and tree trunks and it's difficult to find a model that doesn't have severe cosmetic damage. In recent years, fiberglass reproduction fenders have been manufactured to assist in the completion of restorations. *Jeff Hackett*

The steering wheel and seat were lowered, and the foot pedals were moved rearward. The exhaust was routed out the side, and on later models, the exhaust was routed downward through the fenders. Orchard models are highly sought after by collectors. *Jeff Hackett*

In the Fall 1969 issue of *Better Farming*, Oliver talks about the "Good Deal Time," when it would offer special prices on combines, plows, and tractors for a short time. The article states, "From now until September 30, 1969, Good Deal tractor Specials include Oliver's best-selling models for several hundred dollars less."

Sales of the Specials evidently didn't meet Oliver's expectations. It is said that even though the Specials were stripped-down models, dealers were fitting them with whatever options were available just to get rid of them at the end of the sale. Several Specials have been located with cast wheels, three speeds, and other options that should not have been on that model. Ads were still being run for the Specials in the Winter 1969 issue of *Better Farming*; the 55 series was introduced in November, however, indicating that the 50s were on their way out.

Purple Olivers

Oliver introduced a new sales campaign in 1966 in order to appeal to new customers. The purpose of this campaign was not to sell tractors but to push the superior plows to all farmers. This program was called the Color-Blind Plow.

Oliver boasted that it didn't make any difference whether your tractor was red, green, orange, blue, or even purple. Oliver had a plow that would work great behind a machine of any color. Oliver also wanted to show how fast you could plow with one of these plows.

At the Growing "O" shows that traveled around the country, Oliver included a purple Oliver tractor to pull the plow it was promoting. When the show was held in Mason City and Charles City, Iowa, the driver of the tractor was adorned in a silk jockey suit to demonstrate the speed at which a farmer could plow with this implement. The show consisted of a line of tractors that would take off together and plow a field. The purple tractor was hidden behind hay bales stacked near the starting line and would appear after the other tractors had taken off. Before the other tractors reached the end, the purple tractor would have passed the others like a racehorse. The limited numbers of purple tractors used for the demonstration were returned to the standard Meadow Green color after the shows ended.

The Ski Slope Special

The H.F. Davis Tractor Company of Boston, Massachusetts, conceived the Ski Slope Special, a tractor based on the OC-4 with 60- or 68-inch gauge

Only a handful of tractors were painted purple, but they were not soon forgotten. These tractors were painted purple for advertising purposes and then repainted green after the demonstrations were over. *Author's collection*

Ski slope operators often bolted wide planks onto their tracks to keep the machine afloat. Snow often packed in the tracks creating problems, so the H.F. Davis Tractor Company worked with Oliver to create the Ski Slope Special, using 30-inch-wide steel grousers. *Author's collection*

tracks. The H.F. Davis Company worked with Oliver and the ski resort maintenance crew to build a machine useful for the specific purpose of grooming ski slopes and trails. The Ski Slope Special had 30- or 36-inch steel grouser pads for flotation. This model could be equipped with a wide dozer blade or a modified Model 46 loader unit with a wide bucket.

Levy Front-Wheel Assist

Oliver introduced the hydraulic front-wheel assist, a new four-wheel-drive system, in 1965. This Levy-built unit was first offered on the 1650 and 1850. It was later offered on the 1750, 1950, and 1950-T. Oliver already offered a mechanical four-wheel-drive unit, but hydraulic front assist had advantages such as crop clearance. This axle allowed 26 inches of clearance, which was ample for most row-crop work. The axle was also adjustable for various row widths. Front tread could be adjusted from 62 to 88 inches to accommodate 30- to 40-inch row spacing.

The HFWA system had a much shorter turning radius than a mechanical front-wheel assist unit, largely due to the lack of universal joints interfering. Unlike the mechanical front-wheel assist, there was no skidding of the front wheels. The power wheels were mounted to the axles with a conventional knuckle and spindle assembly. On a 1650 with 62-inch tread and the front wheels disengaged, the turning radius was 203 inches, while it was 183 inches with the front wheels engaged. The wheels did not need to be disengaged when making turns. The tractor actually turned shorter with the wheels engaged.

Tractors that were ordered with the Levy front-wheel assist came from the factory with that configuration. Conversion kits were available to the farmer. In later years, Levy marketed its kits directly to the farmer and for many models besides Oliver. *Author's collection* (left) *Jeff Hackett* (right)

Another advantage of HFWA over MFWA was the ability to synchronize the different tire sizes. Since the front end operated separately from the back end, the speed could be varied to match what the back was doing or to give it a little extra pull. This also added extra traction and lessened the compaction from the rear axle. This option wasn't possible with the other systems.

Most of the farms that required four-wheel-drive power at the time were large operators with large power requirements. This new system had quite a bit of versatility in size and adaptability for smaller farms. It also provided a better ride than the mechanical system.

It was very simple to engage the system. A control lever with six notches for various torque settings was located on the right side of the console. Each of these settings pre-selected the torque delivered to the front wheels. There was also an override setting to obtain maximum torque for short periods in extra-tough spots. When the tractor was under full load with the front end engaged, slippage was prevented. If the tires spun, what seemed like extra power was actually a loss of power being used to shear the soil instead of propelling the tractor forward.

Loader Special

The Loader Special concept was the brainchild of Charley Anderson of Edgeley, North Dakota. Charley was a veteran Oliver dealer in southeast North Dakota. The trend during the 1970s was turning toward the higher-horsepower, high-dollar tractors. Charley, operating as Anderson Brothers, received numerous requests from his customers for a good used tractor. Farmers wanted a reliable gas tractor with power steering and live PTO that a loader could be permanently attached.

Charley went to Oliver and requested it bring back a stripped-down version of the 770. Oliver didn't want to do that but said that if dealers ordered enough of them, the company would design a 1555 to meet their specifications. Charley and his son, Dennis, and Bill Douglas, Oliver's Fargo branch manager, headed to Charles City and spent a week with the engineers to design what would become the Loader Special.

This 1555 Special is fitted with a New Idea loader. However, most of the Loader Specials were fitted with Farmhand loaders. *Jeff Hackett*

The tractor was basically a bare-bones model with no hydraulics or three-point hitch. It had an upholstered seat, but it was mounted rigid on the tractor. The model had 7.50 x 16 6-ply tires on the front and 15.5 x 38 6-ply tires on the rear with pressed-steel wheels. It had a heavy-duty drawbar, three-position tilt and telescope steering wheel, fenders, lights, and an adjustable, wide front axle. The engine was rated at 53 horsepower. The tractor sold for $3,995 with those minimal options. The base price on a gas 1555 in 1973 was $7,325, so the stripped-down model represented a significant savings.

Oliver agreed to build these models if twenty-five of them were purchased at a time. In order to sell twenty-five tractors quickly and recoup its investment, Anderson Brothers began an advertising campaign. This brought about much skepticism by other dealers who convinced farmers this had to be a scam. Anderson Brothers sent out letters to explain how this tractor could be sold so cheap—it was no gimmick, and there was no catch.

The customers loved the new tractors, and they sold like hotcakes. The first order was for twenty-five, the second was for fifty, and subsequent orders led to total sales of 425. Anderson Brothers advertised in a dairy magazine, which appealed to a new audience. Tractors were sold as far away as Powell, Wyoming, and Bozeman, Montana.

Anderson Brothers would send three semi loads of wheat to Minneapolis every week. The trailers would return to South Dakota with 1555s from the Fargo branch and arrive back in Edgeley with nine tractors.

Only after careful examination can you determine if a tractor is a Loader Special. At first glance, the tractor might look like any other 1555, but the drawbar and stationary seat are good indications of this model. Most of these machines spent their lives on dairy farms in the northern part of the country. *Jeff Hackett*

The tractors often sold just as quickly as they were unloaded. Dennis stated that at one time, fifty of them were lined up in front of the dealership. However, due to their popularity, they didn't sit on the lot for long.

Anderson Brothers was also a Farmhand dealer and planned to mount a Farmhand F-10 or F-11 loader on the tractor when it got to the shop. The hydraulics came from a PTO-driven pump or a mounted hydraulic pump. The tractors were also set up for Koyker and Easy-On loaders.

While the tractors were designed to be basic, they could be ordered with factory hydraulics for $4,450. The addition of a three-point hitch cost $4,775, and there was also the option of diesel for $600 more. The tractor had six forward speeds and two reverse but could be ordered with Hydra-Power for an additional cost.

These tractors were first sold exclusively through Anderson. When Oliver saw how popular they were, however, the company began to offer the same deal to other Oliver dealers operating out of the Fargo branch. Williston Farm Equipment was one of the dealers that took advantage of the Loader Special. Dan Baker, the dealer there, said that he ordered five lots of twenty-five tractors during a nine-month span. Many of his customers told him afterward that the only mistake they made was that they should have bought two of them. In fact, several of them did. Dan said that most of his advertising for these models was done on TV and radio. Oliver didn't publish sales literature for the Loader Special.

The Farmer's Union was another business that purchased quite a few Loader Specials. Probably several more businesses purchased the tractors, but it would have taken a good-sized dealership to order twenty-five tractors at one time. Two dealerships alone sold 550 of them from 1972 to 1974. At this time, it is unknown how many were produced in total, but carloads of them were brought into the branch, according to the Fargo service manager.

The Twin 880

In 1959, Oliver's experimental department designed a tractor that would catch everyone's attention. It consisted of two Oliver 880 Industrials joined side-by-by side to make one machine. To make this odd-looking machine even more unique, it was set up to run by remote control. This system was created with the help of an ag engineer from the University of Nebraska, James D. Pichon. This wasn't Pichon's first attempt at a project like this as he also converted an Oliver 770 to be radio-controlled in 1958.

Oliver never intended to build the Twin 880 as a production tractor. It was built as a showstopper and gave the experimental department a chance to play with a futuristic concept of remote-controlled equipment.

The Twin 880 was used for demonstrations at many shows and parades. It caught a great deal of attention when it rolled down the streets during a parade with no driver in site. The operators would walk behind the machine with a radio-controlled device and let the tractor perform while the crowd watched in amazement.

This particular Twin 880 was demonstrated in Waterloo, Iowa, in October 1959. Seated on the tractor is the author's uncle, Leonard Schaefer. While reproductions of this model have been built, none of the actual tractors are still in existence. *Author's collection*

Herb Morrell was Oliver's chief engineer during the XO project. This futuristic model was tested using a four-cylinder overhead valve Hercules engine with 199 cubic inches. It operated at a maximum corrected belt horsepower of 60.9. *Jeff Hackett*

The tractor received more attention at the Oliver dealer show in Waterloo during December 1959 when it was shown in the school gymnasium to hundreds of dealers. The operator reversed one transmission while another went forward to make the tractor spin around in circles. It was a unique demonstration, but it left black skid marks on the gym floor, which the school administrators did not appreciate.

Stories vary as to how many Twin 880s were built for demonstration purposes. According to one former engineer, six were assembled. At least one of them was sold to a racetrack in California, and a couple more were possibly sold for construction. The remaining units were scrapped, and the parts were used to assemble other 880 models.

XO-121

When it comes to the experimental Oliver 12:1 compression, the designation XO-121 says it all. Oliver teamed up with the Ethyl Corporation in the mid-1950s to experiment with a tractor engine capable of burning the high-octane fuels of the future. The companies never intended to build this engine, though. It was an experiment and research tool; thus, the project was called "Ahead of Tomorrow."

Oliver and Ethyl chose to use a Hercules four-cylinder block for the engine. A high compression head was designed with a combustion chamber configuration that was eventually introduced in the 1800 tractor in 1960.

Oliver had learned in the past that compression ratio was the key to turning high-octane fuel into engine efficiency. Tests showed that when the XO-121 went up against the 77, horsepower was increased 44 percent, while fuel economy improved 28 percent.

The tractor's chassis was that of an 88 but used the open-engine style of the Super series tractors, which weren't in production yet. A special chromed grille with vertical bars was used, and Oliver's colors were reversed to give the tractor a red body and green wheels.

When Oliver was done with its testing, this tractor was loaned to a university and then to a living history farm. It eventually made its way back to Charles City and is on display at the Floyd County Historical Society.

Jeff Hackett

Jeff Hackett

Built in Charles City, Iowa, the Massey 98, based on the Oliver 990, was in production for two years, ending in 1960. Massey then turned to Minneapolis-Moline to produce the Massey 97. *Jeff Hackett*

The Massey 98 Tractor

During the late 1950s, Massey Ferguson was looking for a large tractor to help it compete in the high-horsepower market. The company turned to other manufacturers to build the tractors because Massey didn't want to fall behind while it spent time designing and tooling up to build its own models.

Oliver took its successful 990 model with the 3:71 GM engine and converted it to a Massey. The changes made were limited to sheet metal and color with a tag that stated, "Built for Massey-Ferguson by the Oliver Corporation." The familiar Meadow Green and Clover White were replaced by Massey's silver and red color scheme. The hood and grille were styled to match Massey's current line of tractors to give it a look like "one of the family." Initially, the company ordered 500 units from Oliver. While rumors circulated that only 250 were built, records now show that 503 units were actually produced, three of them being prototypes.

HART-PARR
15

CHAPTER 7

OLIVER ACROSS THE POND

Oliver Chilled Plow Works' entry into the export market began in 1872 when it sent a shipment of implements to Japan. Hart-Parr's export business began in 1908 as it exported tractors to Cuba, Chile, Austria, the Philippine Islands, and Argentina. The export market for both companies grew rapidly to encompass a worldwide customer base.

Oliver's plows became so successful abroad that the company laid its claim as "Plowmaker to the World." In 1907, South Africa placed one order for 8,800 sulky plows alone. In 1917, the company was forced to erect a "foreign packing building" in South Bend just to prepare orders for shipping.

Hart-Parr also shipped worldwide but experienced more financial difficulties than Oliver. Tractors were often shipped out on credit, and Hart-Parr experienced difficulties collecting on debts. This created a hardship for the company that it struggled with for years.

There was minimal difference between the domestic and international Hart-Parr tractors. One model in particular, however, was identified as the New Zealand Special. Hart-Parr's largest distributor in New Zealand was Andrews and Beavan, Ltd. of Christchurch. The company found it difficult to sell the Hart-Parr models against the competition because the Hart-Parr model only had two gears. This concern was expressed to management with the request that a third gear be incorporated into the tractor. Hart-Parr agreed to do this if a significant quantity of tractors were ordered. Thus the 12-24, 18-36, and 28-50 New Zealand Specials were born. The tractors are easily identified by the name "New Zealand Special" cast into the radiator. This same configuration was also built as the Australian Special. The Australian model

It was not uncommon to see complete trainloads of tractors being shipped to Canada in the early 1900s. This load of 22-45 Hart-Parrs was being shipped to the Chapin Company, which was a Hart-Parr distributor in Alberta, Canada. Each time a flat car of tractors was shipped, Hart-Parr made up signs with its name on it and the name of the dealer where the train was destined. This picture was taken on November 23, 1909. *Author's collection*

Tractors that were sent overseas were crated up in a different manner than those delivered to Canada. This shipping crew is building a crate so the entire machine can be lifted onto a ship. Notice the front wheels have been removed, the axle stubs are mounted through the timber frame, and the lower rails have hooks for lifting. *Author's collection*

had a special chaff screen on the radiator, and the name "Australian Special" was cast into the radiator.

The Oliver Chilled Plow Works' export division operated within the company. Hart-Parr established a subsidiary in Canada known as Hart-Parr Ltd.

The New Zealand Special is easily identified by the radiator casting. It was built in Charles City, Iowa, and hundreds of these models were shipped overseas. *Jeff Hackett*

When the companies merged, all companies were grouped as one, and the export division operated out of Chicago.

Oliver International S.A./ Oliver Australasia Pty. Ltd.

In 1956, Oliver established a new subsidiary owned wholly by the company to be called Oliver Internationals, S.A. (Sociedad Anonima). This organization was set up with the purpose of selling outboard motors and farm and industrial equipment throughout the world outside of the United States and Canada. The two offices for this subsidiary were located in Venezuela and Chicago. Sam White Jr. was elected president of the organization.

Oliver established another subsidiary known as Oliver Australasia Pty. Ltd. The company had a great market for its crawlers in Australia, as indicated by an order for twelve hundred crawlers placed in 1952. Leslie Dyke, the former export manager in Chicago, was elected president of the new subsidiary.

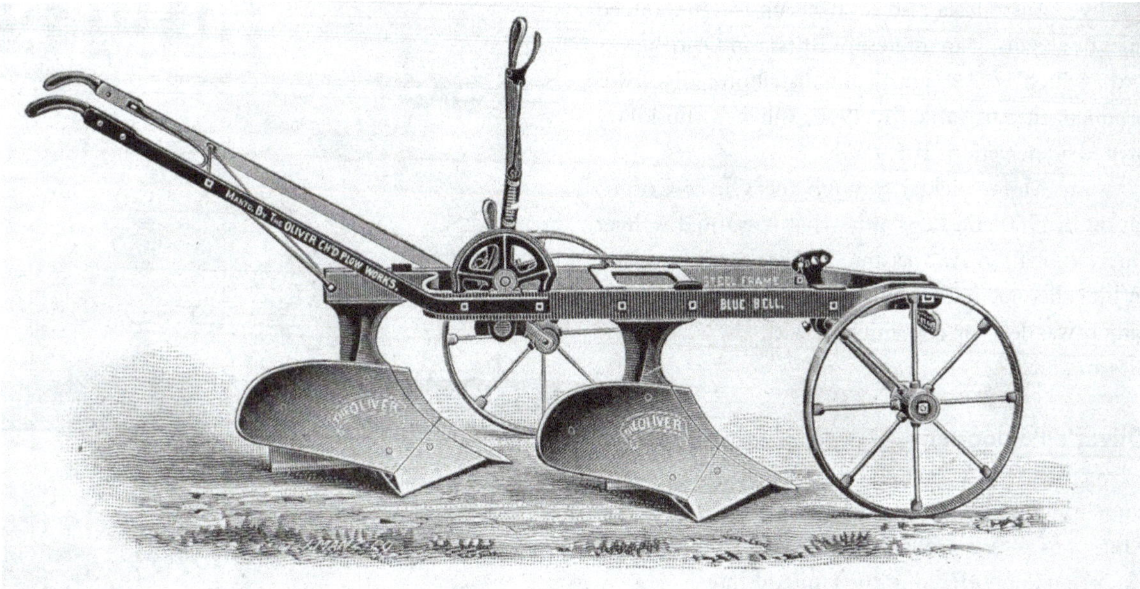

While most plows were designed for horsepower in the United States, many countries were still using oxen or other exotic forms of power. Soil conditions were also different, so Oliver designed other models to suit those needs. This model is called the Blue Bell and was built for Scottish and Australian use. *Author's collection*

This Hart-Parr 16-30 was used in Western Australia for breaking Mallee scrub. It was rolled down with a heavy primitive roller, then burned and plowed. It was a job not possible before the presence of the tractor. *Author's collection*

Oliver set up a plant in Sydney to act as the supplier for the area. Import tariffs were high on complete units, but significant savings could be had if products were brought into the country and assembled there. In order for the serial tag to say "Made In Australia," a certain amount of components had to be built by Australian manufactures. This created a problem because increments were in inches on the US-built parts, while Australia used the metric system.

The initial plan was for Oliver Australasia to produce the OC-12, OC-15, and later crawlers. The manufacture of Australian components was also supposed to increase. The OC-12, however, was the only model actually assembled at the Sydney plant.

Oliver Australasia also set itself up for guaranteed sales by acquiring an interest in Britstand Distributors Ltd., which was a significant distributor of Oliver products in Australia. By 1960, Oliver's subsidiary owned 50 percent of Britstand.

White Motor picked through the Oliver Corporation in 1960 and kept only what it wanted. Oliver International, S.A., became a "keeper;" however, White did not retain Oliver Australasia Pty. Ltd., which was left for the vultures with the rest of the bits and pieces.

Oliver's Payloader

Oliver was well known as an exporter, but it really came to life in 1953 under the guidance of Les Dyke. Many of Oliver's products were built exclusively for export and not offered in the United States.

In 1956, Oliver's export division was looking for a rubber-tired heavy-duty loader to accompany its already successful crawler line. Instead of the costly job of designing and tooling up a machine, Oliver turned to the Thew Shovel Company of Lorain, Ohio.

Thew had its own payloader line known as the Thew Moto-Loader. Oliver marketed the ML-153 loader as the Oliver OL-175. The gas version of this machine was powered by a Hercules GO-339 engine, while the diesel version was powered by a Hercules DRXB engine. The OL-175 had a 1 3/4-yard bucket (early literature shows it featured a 1 1/2-yard bucket) and full reversing Torqmatic transmission. The OL-175 is pictured in a 1956 issue of *Oliver Export News* although no number designation had been assigned at that time.

Competition was stiff in the payloader business, and Oliver was up against the Hough HH, Michigan 125-A, Allis-Chalmers TL-20D, and the Yale 154 Trojan. Most of these advertised a two-yard bucket. It didn't take long for Oliver to realize it needed to keep up, so the OL-175 was upgraded to the OL-200 in 1958. The OL-200 was basically the Thew ML-157.

The OL-200 was very similar to the OL-175. Both models used the same Hercules gas and diesel engines as the 175. The OL-200 also offered an optional Cummins JF-6-B1 diesel engine. The gas engine was rated at 109 horsepower, the Hercules diesel was 133 horsepower, and the Cummins was 150 horsepower.

The OL-200 payloader was available in 1958 for $19,600 when fitted with the Hercules diesel engine. The operating weight was 20,000 pounds. This model competed with the Michigan 125-A and the Allis-Chalmers TL-20 D. A few OL-200 models have been located in the United States. While they are identical to the Thew ML-157, they can be easily identified by the Oliver name cast into the rear weight and the Oliver serial number plate located inside the cab. *Author's collection*

The Cummins diesel engine was cheaper than the Hercules diesel.

The OL-175 had a torque converter ratio of 2.5:1, but the OL-200 was 3.5:1. The OL-175 could be ordered with 14.00 x 20 or 14.00 x 24 12-ply tires, while the OL-200 was only offered with 14.00 x 24 tires. Both machines used planetaries and had four-wheel drive.

The peak lift on the OL-175 was 11,000 pounds, while the larger OL-200 was rated to handle 14,000 pounds. For $750, the OL-200 could be ordered with a 3-cubic-yard light material bucket.

The top speed for both models with the 24-inch tires was 24 miles per hour, and low range was 3 miles per hour. Both models were available with a cab for an extra $495.

The OL loaders boasted that they had a low center of gravity, even weight distribution, and balanced power. The tractors were advertised as an "all around machine that gives peak performance." They were capable of carrying a full load up a 30-degree incline. Features included power steering and hydraulically operated brakes with vacuum boosters.

Oliver's export division was not very successful in selling Oliver payloaders. The company eventually offered the loaders for sale through the domestic industrial division out of Cleveland. This division didn't push the sales of them either, and the industrial sales force was very weak. By 1960, the payloader line was dropped.

The MRS Tractor

In 1955, Oliver teamed up with Mississippi Road Service (MRS) to sell four models for the export market. This industrial tractor was designed to pull heavy loads at a higher speed. The smallest of the four models was the Model 125 with 210 horsepower. The largest in the lineup was the Model 200 with 335 horsepower.

The Oliver/MRS was teamed up to a Woolridge scraper in most applications. Woolridge devised a cylinder mounted at a 45-degree angle that would transfer weight from the scraper to the rear of the tractor for added traction.

Sales of this unit were minimal due to the high cost of a machine of this type. The Oliver/MRS tractors weren't attractive to the buyer even though the machine was equipped with high horsepower. Oliver International only offered the unit for a short period of time, but it was long enough to make it into the pages of the *Oliver Export News*.

Oliver SAME & Arbos

The Oliver farm equipment line was a US company, but unknown opportunities were available overseas. Prior to the White Motors Company buyout, Oliver had an export division, but the equipment sold was US-built by Oliver. Things changed when White Motors came along. It was White's opinion that it was becoming increasingly difficult to keep pace with the growth of foreign markets without foreign assembly or manufacturing plants. Tariffs were too costly to import products from the United States. In 1964, White Motors appointed Hans U. Graf as director of international operations to study the European market.

Even though the SAME tractors were not built for sale in the US, they were painted Meadow Green and Clover White and had a similar styled grille to match the US models. A few examples of this tractor have made it back into the domestic market. *Author's collection*

By 1965, the international division had made great progress in broadening overseas distribution and developing a basis for the manufacture of products in Europe. In 1966, White announced it had acquired a 75 percent interest in Arbos. S.p.A. Arbos, located in Piacenza, Italy, became a subsidiary of White and manufactured combines for rice and grain.

White's 1966 Annual Report reads: "The overseas activities of all divisions have been consolidated. This concentration of effort is designed to improve penetration of international markets. White Motor International, N.V., based in Amsterdam, Netherlands, with a branch in Paris, France, is developing markets in Europe and the Middle East. It is also responsible for the operation of our affiliate, Arbos,

Hart-Parr tractors were used for more than plowing fields and building roads. This 30-60 was used for digging entrenchments for the troops in Vienna, Austria. *Author's collection*

which manufactures combines at the present time. Arbos has plans to expand its line of products to include other farm equipment. Considerable progress was made in establishing new dealers in a number of countries during 1966."

White acquired the remaining 25 percent of the minority interest of Arbos in July 1967, bringing its ownership of the company to 100 percent. Arbos had excellent sales in the Italian, African, and South American markets. In 1966, it was ranked fifth in sales in the Italian market and was first in sales for 1967. Arbos made good progress in the export market and held a dominant position in Portugal. It was also well established in Venezuela, Sudan, Spain, and other countries. Arbos increased production by 35 percent in 1968, and arrangements were also made to start assembly of Oliver Arbos combines in Turkey.

In 1970, White decided a tractor needed to be added to the equipment line for its international dealers to sell. White turned to one of the most reliable and well-known manufacturers in Italy: SAME (pronounced Sommy).

The SAME company started in the 1920s when a young man named Francesco Cassani started to tinker with the idea of an engine that burned something besides gasoline. He succeeded in 1928 and constructed the world's first diesel-powered tractor. Cassani's mother used the money she had

The Arbos Oliver and White combines were built as hillside models and some were offered with tracks in place of wheels. The hillside models were powered by Perkins diesel engines and had the ability to hydraulically tilt 40 degrees in order to keep the machine level on the slopes. *Author's collection*

SAME, an Italian tractor manufacturer, is credited for building the world's first diesel powered tractor in 1928. It was a successful company, which helped Oliver market tractors for the export market. *Jeff Hackett*

from running an ice-making factory to finance a building for the machine's production.

While Cassani struggled with distribution and marketing problems, Caterpillar allocated nearly its entire engineering and research budget to work on its diesel project. After spending more than $1 million, Caterpillar introduced the Model 60 that burned diesel, or commercial fuel oil, as it was called in those days. By 1931, Cat had a diesel engine on the market and took the glory away from Cassani.

Cassani set out to perfect his diesel engine and started building engines for trucks, boats, tractors, and aircraft. In 1942, Cassani and his brothers formed Societa Anonima Motori Endotermici (SAME), meaning "internal combustion engine corporation." The Cassani brothers opened a new shop and set out to build a new tractor. They succeeded and introduced the first four-wheel-drive diesel tractor to the world in 1952. Francesco Cassani had dreamed of this concept ever since he saw Jeeps crawling across all the rough terrain during the war. It took quite a while to convince the farmer that four-wheel drive was the wave of the future, but the company began to grow once the idea took hold. The unit was very reliable and allowed a decent turning radius.

Francesco Cassani was a true inventor. He lived to build and design and came up with many innovations in use today, such as the electronic power-shift transmission that is patented by SAME. In time, the company merged with car manufacturer Lamborghini and Hurlimann.

When White turned to SAME, it was producing models from 46 to 85 horsepower. SAME took the models that it was building for itself and painted them green and white, changed the grille, and added the Oliver/White decals without making any mechanical changes. The first two digits in the model number designated the horsepower, and the last digit was the drive (two or four). Thus a 452 model would have 45 horsepower and have two-wheel drive. Ironically, this is the same manner of model designation that was used on the Oliver Mighty-Tow line in the late 1960s. It was carried over into the White line for many years.

The new Oliver models made their debut in Paris during the spring of 1970. The 462, 464, 562, 564, 672, 674, 852, and 854 were available. SAME updated its tractor program in 1972, which also caused changes in the Oliver/White lineup. The models now available included the 452, 454, 562, 564, 682, 684, 782, 784, and the bigger 982 and 984 models. The comparable SAME models are listed on the following pages with the Oliver designation.

The SAME/Oliver connection continued through 1975, but during that time, Italy was hit with a tough

The 782 and 784 models were rated at 78 horsepower and powered by a four-cylinder, 254-cubic-inch air-cooled diesel engine. With a 12-speed transmission and front-wheel assist, this model was good for tough field conditions. The downward exhaust was optional, and most were fitted with the standard vertical muffler. *Author's collection*

recession and the line's value decreased. The combine factory and the rest of the farm equipment line suffered as well. White owned 100 percent of the combine plant and decided to cut its losses and shut the door on its Italian venture. The company was already experiencing its own hardships, and this failed Italian venture didn't help. White sold the Arbos plant, product line, and inventory in 1976. With no combine to sell, White also ended its venture with SAME.

El Toro

Oliver attended the 1964 US National Exposition held in Mexico City. The export division was well represented at this demonstration. In order to present a model that would be the icon of a massive animal, Oliver tagged its four-wheel-drive models "El Toro," meaning "the bull." Tractors sent to Latin America and equipped with four-wheel drive had an El Toro decal on the side panels starting in 1964. There was nothing unique about the tractors other than four-wheel drive. The El Toro model came in three different sizes: 1900, 1800, and 1600. These models, with the four-wheel assist, delivered 30 to 40 percent more drawbar pull than models of comparable size with rear-wheel drive only.

This El Toro model has been fitted with optional Terra Tires for better flotation. While most El Toros were fitted with rice tires for better clawing power, this variation worked well for staying on top of the ground and has become highly sought after in the collector world. *Super T*

Sales literature from the 1964 *Oliver World* colorfully describes the models:

"When a beast of burden is the prime mover, the one prized most leans into the traces and digs in with all four feet instinctively and eagerly. But the modern agriculturist has found that mechanical brute, the tractor, is far superior to any draft animal.

"Now there's a purring green giant of a new type that pulls more than hundreds of hooves by gripping the ground with all four wheels and hanging on with dozens of rubber claws.

"It's the Oliver El Toro four-wheel drive built in three sizes to provide the exact power needed in any farm operation. Biggest of all is the mighty 102-horsepower 1900 diesel with a maximum drawbar pull of 16,900 pounds. Next is the record-breaking gasoline or diesel 1800 with more than 80 horsepower that pulls up to 14,000 pounds on the drawbar. Completing the broad Oliver line is the new 60-horsepower gasoline or diesel 1600 that develops 11,000 pounds of pull at the drawbar."

It was obvious what the real bull was after operating all three models. When the 50 series appeared, the 1950 was the only El Toro model. It didn't need Terra tires to be considered an El Toro. All it needed was four-wheel drive.

The first El Toro models were introduced in 1964 with the 100 series tractors. Oliver didn't introduce the Goodyear Terra tires for sale until July 1965. Terra tires weren't even offered when the El Toro was introduced.

To make the El Toro even more rare, this model was sent to Latin America only through Oliver International, S.A. No tractors tagged "El Toro" were sold in the domestic market, and nothing on the build card indicated the tractor was tagged an El Toro. If the tractor was built for export, however, the destination would have been marked Norfolk, Virginia. This was the shipping port where equipment was loaded on the boat and sent downstream to Latin America.

When tractors were exported, they had to be equipped with tires that could be serviced in the export country. Goodyear had many dealers in Colombia, so the entire fleet that went to Latin America was equipped with Goodyear tires.

This original export photo shows an entire shipment of El Toro models being delivered to the Bavarian Sugar Company. These models were quite popular as they could easily pull the heavily loaded wagons long distances to the sugar mills. *Author's collection*

HANDBUCH

FÜR

HART-PARR TRAKTOREN

TYPEN

12-24

18-36

28-50

HERSTELLER:

OLIVER FARM EQUIPMENT COMPANY

GRUENDER DER TRAKTORENINDUSTRIE

CHARLES CITY, IOWA, U.S.A.

EUROPAEISCHES VERKAUFSBUREAU UND TRANSITLAGER:

91, RUE LOZANE, ANTWERPEN

BELGIEN

Parts manuals and operators books were often printed in different languages, depending on where the machine was going. *Author's collection*

CHAPTER 8

MISCELLANEOUS PRODUCTS

In 1924, Hart-Parr, the company that made its mark on the tractor industry, began selling two models of washing machines. The first was the Model A, which had four legs. The Model B had three legs. Both models were priced at $155, used a copper tub, and were operated with electricity. The tub rotated within the frame while plungers inside the tub moved up and down to pound the dirt out of the clothes. They came equipped with a Westinghouse electric motor or a Cushman gas engine. The Hart-Parr machines were expensive models, and the company faced stiff competition from other manufacturers. This venture proved to be unsuccessful and ended in 1927.

Railroad Tools

The Tallman family owned the Shelbyville, Illinois, plant prior to Oliver's purchase and had gained much success with their line of hay equipment. The Tallman family was also in the business of railroad tools used to build and maintain tracks.

Tools built for rail construction were numerous and ranged from picks to dig ground for proper tie placement to track levels to make sure the rails were even. There were also screw-type rail punches and rail benders. An Oliver catalog shows eleven different models of levels; most of them were built out of thoroughly seasoned western pine. Each of the tools was identified either by the name "Tallman" or "Oliver" after Oliver acquired Tallman. The name "Oliver" is stamped into the pick's head.

The railroad tool division was eventually sold by Oliver to the Gibraltar Equipment & Mfg. Co. of Alton, Illinois. It began to manufacture the tools as GEMCO tools. The company disappeared in 1965.

The Gibraltar Equipment letterhead calls the company "Successors to Tallman Manufacturing Company

The design for the Hart-Parr washing machine was purchased from E.F. Beebe of Minneapolis, Minnesota, who held the patent on his invention. Hart-Parr manufactured these machines in its plant to replace the void in tractor production during the depression.
Author's collection

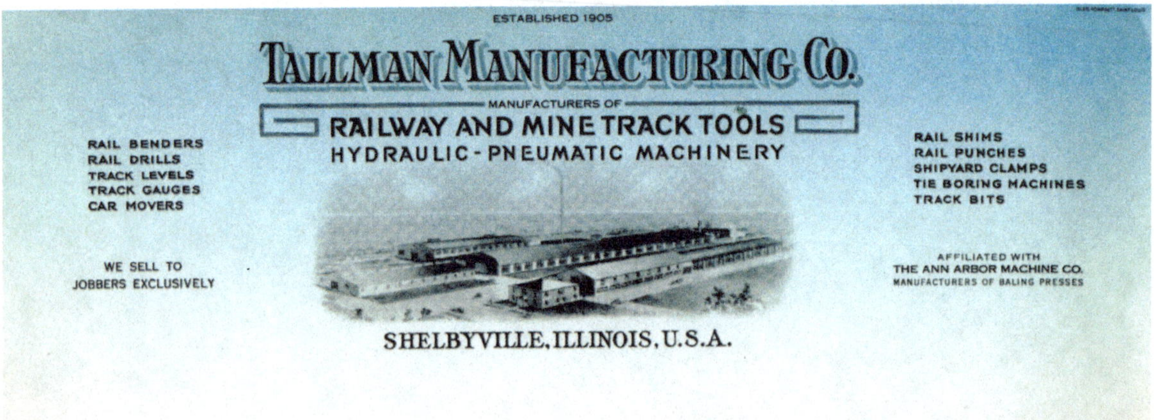

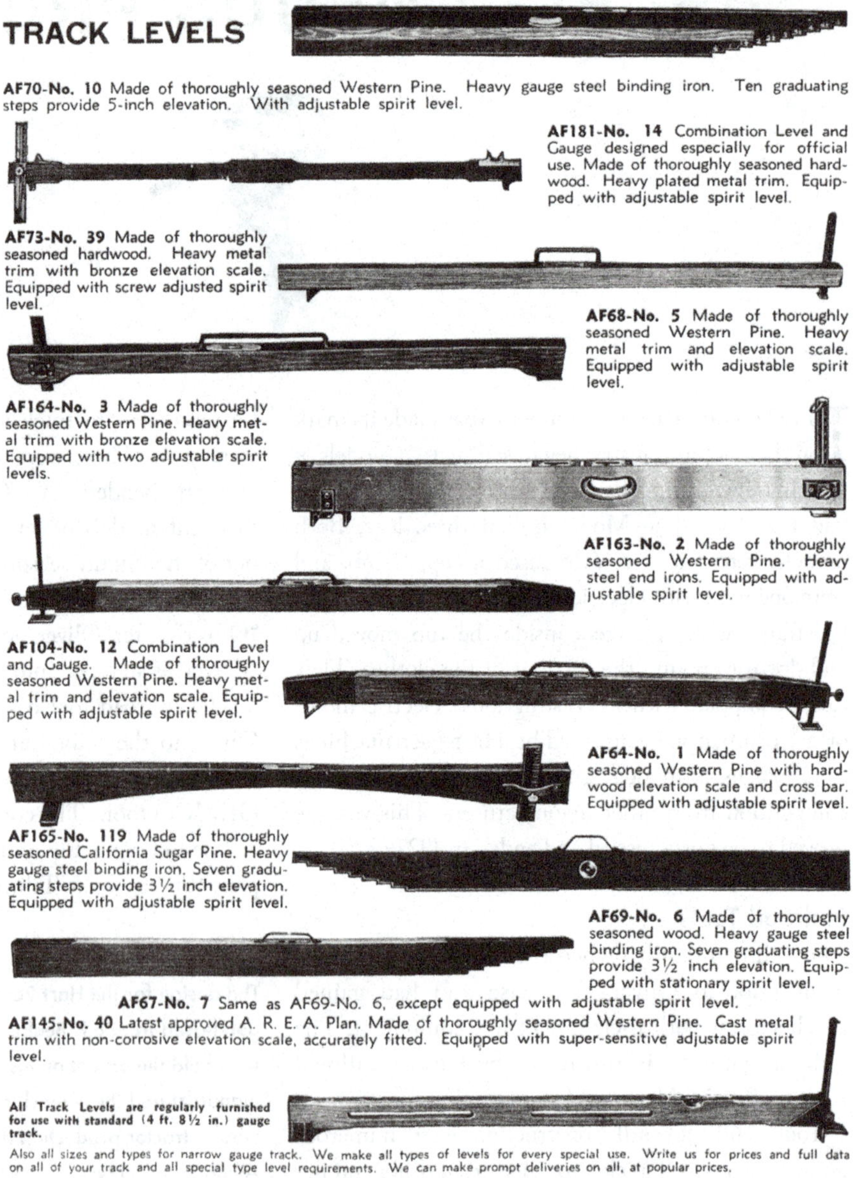

TRACK LEVELS

AF70-No. 10 Made of thoroughly seasoned Western Pine. Heavy gauge steel binding iron. Ten graduating steps provide 5-inch elevation. With adjustable spirit level.

AF181-No. 14 Combination Level and Gauge designed especially for official use. Made of thoroughly seasoned hardwood. Heavy plated metal trim. Equipped with adjustable spirit level.

AF73-No. 39 Made of thoroughly seasoned hardwood. Heavy metal trim with bronze elevation scale. Equipped with screw adjusted spirit level.

AF68-No. 5 Made of thoroughly seasoned Western Pine. Heavy metal trim and elevation scale. Equipped with adjustable spirit level.

AF164-No. 3 Made of thoroughly seasoned Western Pine. Heavy metal trim with bronze elevation scale. Equipped with two adjustable spirit levels.

AF163-No. 2 Made of thoroughly seasoned Western Pine. Heavy steel end irons. Equipped with adjustable spirit level.

AF104-No. 12 Combination Level and Gauge. Made of thoroughly seasoned Western Pine. Heavy metal trim and elevation scale. Equipped with adjustable spirit level.

AF64-No. 1 Made of thoroughly seasoned Western Pine with hardwood elevation scale and cross bar. Equipped with adjustable spirit level.

AF165-No. 119 Made of thoroughly seasoned California Sugar Pine. Heavy gauge steel binding iron. Seven graduating steps provide 3½ inch elevation. Equipped with adjustable spirit level.

AF69-No. 6 Made of thoroughly seasoned wood. Heavy gauge steel binding iron. Seven graduating steps provide 3½ inch elevation. Equipped with stationary spirit level.

AF67-No. 7 Same as AF69-No. 6, except equipped with adjustable spirit level.

AF145-No. 40 Latest approved A. R. E. A. Plan. Made of thoroughly seasoned Western Pine. Cast metal trim with non-corrosive elevation scale, accurately fitted. Equipped with super-sensitive adjustable spirit level.

All Track Levels are regularly furnished for use with standard (4 ft. 8½ in.) gauge track.

Also all sizes and types for narrow gauge track. We make all types of levels for every special use. Write us for prices and full data on all of your track and all special type level requirements. We can make prompt deliveries on all, at popular prices.

The Tallman Manufacturing Company, which was acquired by Oliver, had a railroad line right outside its back door. While the company was in the farm equipment business, there had to have been some personal connection with the railroad lines for it to start manufacturing the wide variety of tools that it made to serve that industry. In the Oliver collecting world, the railroad tools are often overlooked.

Author's collection

Oliver had been selling power units for many years, but they were primarily for irrigation, pumping, or generator units. Since the company was diving in the water with its outboard line, the marine power unit seemed like a good fit, but it was a hard sell for a farm equipment company. *Author's collection*

and the Track Tool Division of the Oliver Corporation." There is little information available on this line of equipment through Oliver or Gibraltar.

Oliver Marine Engine

Oliver sold a marine engine produced by Waukesha. Harry Horning, founder of Waukesha Motors, got his start in the motor business with marine-type engines. While the marine engine sold by Oliver was basically the same as the ag engine, it had a water-cooled exhaust manifold and a deeper oil pan to allow the oil to sway in rough waters.

When Oliver started to market the marine engines, they were referred to as "work boat diesel engines" in price sheets and stocked at South Bend Plant No. 2. When the plant shut down in early 1958, the stock was sent to Charles City and orders were placed through that plant.

Regular equipment on the work boat engines was air cleaner, fuel supply pump, injection pump, variable-speed governor, oil filter, water pump, fuel filters, flywheel and housing, 12-volt generator, solenoid starter, starter controls, and thermostat. Extra equipment consisted of an instrument panel and electrical instruments, governor control, tachometer, engine front support, reduction gears, and reverse gears. The latter were both provided by Snow-Nabstedt. The sale price of a bare Super 166 boat engine was $1,504, while the Super 199 sold for $2,597. They were rather pricey for the time, and they don't appear to have been very successful since the line was dropped in 1960.

Oliver's Aviation Division

During World War II, almost every manufacturer converted their plants to assist with the war efforts. After the war was over, plants had to switch over to build new products or employees would be laid off. Oliver's Battle Creek division looked for more work, specifically in the defense department, after the war. It was able to fill a void when it formed the Oliver Aviation Division and secured a contract as a subcontractor for the Boeing Airplane Company.

Oliver purchased the former plant of the Goss Printing Press Company in Battle Creek, Michigan, during August 1951. This spacious plant had tall

Few people involved with farm equipment know that Oliver had a hand in building aircraft. The Boeing RB47-E was Oliver's largest project and was quite a prestigious one. *Author's collection*

ceilings and cranes. They were perfect for Oliver, which was anticipating its upcoming contract with Boeing. On November 17, 1951, Oliver received a letter of intent from the Boeing Airplane Company to allow Oliver to build the main fuselage assembly of the RB-47E Stratojet, and the Oliver Aviation Division was formed. The contract that made it official was signed on January 26, 1952.

Oliver had absolutely no experience in aircraft assembly but accepted the responsibility of meeting the strict deadlines and high quality standards set forth by Boeing. Upon the hiring of employees for the aviation division, key people were sent to a five-week training school conducted by Boeing at its plant in Wichita, Kansas.

Training programs began in Battle Creek for the aviation division to properly teach the production personnel and technicians about Boeing's rigid specifications. Over four thousand people went through the training program during the era of Oliver's Aviation Division. Employment reached nearly thirty-nine hundred people during its peak.

Production in the aviation division got off to a rough start, so Oliver hired Red McMinds to be the manufacturing superintendent and to get things whipped into shape. McMinds made charts for everything. He was successful and got the plant up and running efficiently so that all deadlines could be met and could be passed. Working with Boeing was an honor but not an easy chore. Only the best of the best secure contracts with such a prestigious company.

The Oliver aviation plant (also known as Battle Creek Plant No. 2) consisted of 280,000 square feet of floor space situated on 11.4 acres. Although the plant was adaptable for aircraft production, quite a bit of money was spent on renovations and alterations. Oliver provided the land, buildings, a substantial amount of tools, and all office equipment at a cost of $1.68 million.

Even though the contract was signed in January 1952, the first completed fuselage didn't leave the building until August of that year. The fuselage made its first launch into the blue skies over Kansas a year later in August 1953.

Oliver tooled up to build nearly every part used to assemble the fuselage, an endeavor that cost $7.4 million. Oliver employees did all of the welding, riveting, plumbing, and wiring. It was an intricate process in a time when human hands, and not robots, did all the work.

The first contract between Boeing and Oliver called for 52 units. Another contract was signed for an additional 168 units, and the third and last contact called for only 35 units for a total of 255. Of the total production, 240 units were RB-47E models and 15 of the units were RB-47K models built after the RB-47E model. Oliver built all RB-47E aircraft fuselages.

The RB-47E was a photo- and weather-reconnaissance version of the B-47 bomber plane built for the US Air Force. The bomb bays were removed and replaced by photoflash bombs for nighttime aerial photography. It was powered by six General Electric J47-GE-25 jet engines and was manned by a crew consisting of a pilot, co-pilot, and photographer/

Upon Oliver's completion of the fuselage section, it was completely covered to hide its identity and shipped to Boeing in Wichita, Kansas, for final assembly. *Author's collection*

navigator. The RB-47E had a wingspan of 116 feet and was the first jet-powered plane after World War II.

The K model had additional windows in it for side radar. It was used for gathering weather data and sampling air that could be tested for radioactive fallout after nuclear weapons testing.

The B-47 was state-of-the-art in its day. It was the first jet-powered bomber and the first with swept-back wings. Several different variants of this model were built, and two had their Section 42 main fuselages built by Oliver. All sheet metal was installed in the fuselage, and the insides were ready to go. It weighed approximately 6,700 pounds and was 10 feet wide and 45 feet long. Once the fuselage was complete, it was loaded on a special railcar. The rail carrier had to take a special route due to the size of the aircraft. The destination was Boeing in Wichita, Kansas. From there it was unloaded, and the wings and other assemblies were put together to complete the aircraft.

There are no identifying marks on the fuselage to label it as an Oliver. Boeing would allow no names on any of its assemblies. In fact, one brave assembler decided to take a drill bit and carve his name inside of one of his units. Upon the discovery of this fact, his unit had to be disassembled and the name removed. He was reprimanded and surely never tried to pull that stunt again.

The employment situation at the Oliver plant was not without problems. Employees were members of the UAW (United Auto Workers) because of its proximity to Detroit. Several walkouts took place throughout production, and this concerned Boeing. Walkouts meant that there was a possibility that deadlines would be missed. Boeing decided that it would not be renewing its partnership after the third contract expired. Red McMinds heard that the contract would eventually come to an end and decided to leave Oliver in search for more permanent employment. Theron Tallman was promoted to superintendent of manufacturing in February 1954. There was still enough work left to last until the end of the year, when Oliver's contract with Boeing would be brought to a close.

Oliver Outboard Division

When Oliver realized that it was going to lose its livelihood with the aviation division, the company set out to fill the void. It tried for more government contracts but was unsuccessful.

At about this same time, the Chris-Craft Corporation, which had been building boats since the late 1800s, found itself in a predicament of its own. After World War II, it had decided to try its hand at outboard motors and hired several engineers from Mercury, who brought with them some of their designs. Mercury went after Chris-Craft for patent infringement and told the company to "fade completely from the outboard scene or a sizeable lawsuit would be forthcoming."

The best solution for Chris-Craft was to sell off its outboard line and keep producing boats. Oliver purchased the Chris-Craft line of outboard motors, and production began in March 1955 at the building formerly used by Oliver's aviation division in Battle Creek, Michigan.

Oliver engineers took the lower units used on Chris-Craft's Commander and Challenger and redesigned them. These models ranged from 5 1/2 horsepower to 15 horsepower. The outboard line grew during the next few years with the largest models being the Twin

Oliver began to advertise its outboard motors in sporting, boating, and recreational magazines. It was a growing industry, and Oliver offered a wide variety of sizes in hopes of capturing a piece of the pie.
Author's collection

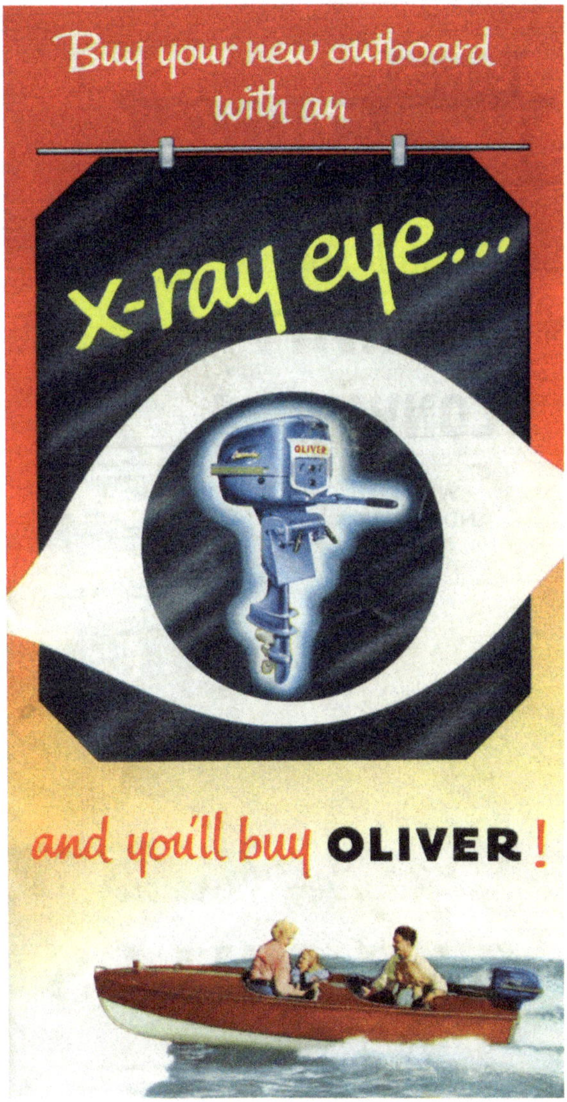

Oliver spared no expense when it came to promoting the outboard motor line. Every ad piece had colorful artistic drawings to mesmerize anyone who looked at it. *Author's collection*

Oliver's advertising department set up a photo shoot that concentrated on family activities. The theme seemed to be, "Have fun with an Oliver!" *Author's collection*

70 with 35-horsepower counter-rotating units. The outboards carried the names of Mohawk, Ranger, Lancer Olympus, and Bulldog.

Oliver supplied outboard motors to the Western Auto hardware store chain in 1957 and 1958. Four different sizes all carried the name of Wizard. An Oliver-built Wizard will always have a serial number that starts with OC (Oliver Corporation).

The outboard motor division did not prove to be profitable, however. The outboards were marketed through the farm equipment dealers and were often given away as door prizes at shows.

In a cost-saving measure, Oliver sold all manufacturing rights and outboard equipment to F. Perkins Ltd. of England in early 1959 for an exchange of stock. Oliver contracted with Perkins to manufacture the 1960 line of motors. They were built to Oliver's specifications and sent back to the United States. This contract went into effect in 1959 but was short-lived. In March, Oliver announced the contract with

The Oliver-Farquhar conveyor systems, built in York, Pennsylvania, can still be found in many post office facilities and grocery distributors. *Author's collection*

Perkins would be terminated on October 31, 1960, due to the inability of Perkins to supply Oliver with the sufficient quantities of motors. Ironically, this was effective the same day White Motors took over Oliver.

Oliver wasn't ready to throw in the towel on the outboard line just yet. It enlisted the help of industrial designer Richard Arbib to provide a concept drawing of the next Oliver outboard model in 1961. The engine never went into production, however, and Perkins continued to manufacture outboards under its own name using the tooling and designs acquired from Oliver.

Cider Presses/Conveyors

The hydraulic cider press is one item in the Oliver-Farquhar line that is often overlooked. This item was brought into the Oliver family with Farquhar's acquisition and continued to be manufactured and sold by Oliver. The company produced five different sizes of presses, ranging from the small model designed for the roadside orchard stand to a commercial model. The small model had a capacity of 600 gallons in a 10-hour time span. The largest commercial model had a capacity of 16,000 gallons per day. These presses were used mostly to make apple cider, but they were also good for vegetables and other fruits.

The conveyor line was another successful product built under the Oliver flag at the Farquhar plant in York, Pennsylvania. Several different models were available, including a trough conveyor that was used for handling coal, stone, and sand. A drag-type conveyor was popular for unloading railcars, and the flat belt conveyor was popular for moving boxes and crates. One of the most successful lines in the later years consisted of the roll-free gravity conveyors.

When Oliver acquired Farquhar, it also took on the line of sawmills offered by the company. The smallest model could be operated with as little as 20 horsepower, while the large commercial model required 150 horsepower. *Author's collection*

Many of these conveyors can be found in the larger post offices. The company had a large government contract to install these conveyors in post offices around the country, and this kept Farquhar's doors open after White purchased Oliver.

Farquhar offered many other products in its lineup, and the sawmill was another popular item. Built in a variety of sizes for the small farm operation to the larger models used in commercial work, the largest model had the potential of 25,000 board feet per day. In 1950, prices ranged from $1,000 for the tractor mounted saw to $3,500 for the 10,000-pound commercial unit.

Town & Country Mowers

Top sales were the name of the game during the 1960s. Companies would market practically any piece of equipment they could sell. Every customer that walked through the dealership had a piece of land with a yard that needed to be mowed. Tractor manufacturers realized if they could sell a customer a tractor, they could most likely sell him a lawn mower too.

Nearly every major tractor manufacturer offered a lawn mower. While they were rarely built by that company, they carried the company's color and name, which was enough for the farmer loyal to his brand.

Minneapolis-Moline began selling a series of Town & Country mowers built by the Jacobsen Company of Racine, Wisconsin, as early as 1962. They were known as the MOCRAFT line and were painted brown. Later, the models took on the familiar Harvest Gold MM paint.

White Motors decided to combine Oliver and MM into the White Farm Equipment Company in late 1969. The MM mower's color changed to Morning Glory Blue and Misty White in 1970. This allowed the line to be sold by both companies, which, although they had been combined into one company, they operated separately and had their own price books. According to the Minneapolis-Moline price book, the mower could still be ordered with the yellow paint and MM decals. The blue and white model was also marketed to the Canadian farmers, as shown in the 1970 Cockshutt full-line catalog.

The 107 was the smallest machine in the lineup and had a 7-horsepower Briggs & Stratton engine. It could be purchased as a 107E with an electric

The Town and Country Series of mowers weren't built for very long but are the ultimate find of every collector. The 125 models were built with a gear transmission or hydrostat. It was the only model built in two configurations. *Jeff Hackett*

The green Town & Country series was built in 1972 only. Transmissions were provided by Peerless, a division of Tecumseh. A limited number of them were built for the Canadian market and badged in the Cockshutt colors and name. *Author's collection*

start or as a 107M with a manual or rope start. The 1972 list price on the 107M was $609. The 107E was an additional $100. The 107M weighed 389 pounds, while the 107E weighed in at 437.

The 108H was the next model of mower. It featured an 8-horsepower Briggs & Stratton engine with hydro drive. Electric start was standard, and the machine listed for $919. The headlight kit was $20 for the pair, and the rear floodlight with flasher was an additional $20.

The 110 mower used a 10-horsepower Kohler engine with a gear drive transmission. It was listed at $988. The 112 was basically the same machine but used a 12-horsepower engine and was available as a gear drive or hydro. The Hydro model listed for $1,248 in 1972.

The 114 was the biggest model in the White Town & Country series. The unit had a 14-horsepower Kohler engine with a hydro drive transmission. The list price for this 688-pound model was $1,488, and quite a few options could be purchased for it such as front and rear wheel weights, chrome hubcaps, tire chains, rear floodlight, PTO, Category O hitch, and a cigarette lighter. A mower with all of these options cost more than $1,710.

In 1972, the company name was changed back to the Oliver Farm Equipment Company, a subsidiary

The 125 Hydro listed for $1,480, while the gear drive model was only $1,285. These were quite pricey for their time and were a tough sell to the farmers. *Jeff Hackett*

of White Motors Corporation. Everything became Oliver. There was only one price book, and the only merchandise that was Minneapolis-Moline was the unsold inventory. There was a lot of unrest among the former MM dealers. Groups of MM dealers got together and went to the home office of White Motors and demanded yellow tractors because they did not want to sell the green equipment. The G550, G650, G750, and G850 were the answer to this problem. The G-955 and G-1355 models were later developed in this same color scheme.

The Town & Country Line changed from the blue and white White model to a new Oliver Meadow Green and Clover White Town & Country for 1972. These models were still being sold in 1973 but apparently leftover models. The machines were built by Jacobson but were a redesigned model with a longer wheelbase, stronger frame, and different sheet metal. While the Oliver T&C was marketed to Canadian farmers as a Cockshutt machine, few were sold due to their high price. The Canadian machine had red sheet metal and Cockshutt decals.

The Oliver T&C was offered in four different sizes and five models: 75, 105, 125G, 125H, and 145. The 75 was the smallest of the series with a 7-horsepower Briggs & Stratton 170 707 engine. It was the only unit of the series powered by a Briggs & Stratton. The transmission had three forward gears and one reverse with a top speed of 6 miles per hour. The list price for the 75 was $760. While the 75 was a smaller, cheaper model, it had an electric start and fuel gauge as standard equipment.

The 105 was powered by a 10-horsepower Kohler K241AS engine. The engine base was actually welded to the uni-frame. The 105 had a four-forward-speed transmission with a top speed of 6.7 miles per hour.

The 125 was built in two different versions. One was the 125G, which had a standard gear transmission. The other, the 125H, was a hydrostatic unit. Both models were powered by a 12-horsepower

The 75 model was the smallest mower in the Oliver line. It was also the only model to be powered by a Briggs & Stratton engine. *Jeff Hackett*

Kohler K301AS engine. The 125G had four forward speeds and one reverse with a top speed of 6.5 miles per hour. It had a 12-volt electrical system with safety lock starting, an alternator, and electric implement engagement with the flip of a switch. Manual lift was standard, but a power lift was available as optional equipment. The weight of the Hydro model was 770 pounds, while the gear model was 750 pounds.

The 145 was the deluxe machine of the series. A 14-horsepower Kohler K321AS engine with balancing internal counterweights powered the unit. A hydrostatic drive transmission was standard, and it also came with hydraulic lift and lights. The 145 had a fully adjustable, cushioned seat. An electromagnetic clutch operated the PTO for the mounted implements. Because the 145 was the most expensive model, it was produced and sold in the lowest quantities.

A 165 model was on the drawing board and planned for sale under the Oliver name. Later service bulletins list the 165 model, but none were actually produced and sold.

White was starting to introduce its Field Boss series of tractors by late 1973. The Jacobsen-built Town & Country tractors, although well built, were too costly for the company to buy and try to sell at retail. White had no choice but to go to a more competitive manufacturer. In October 1973, while dealers were being introduced to the White 4-150, they were also introduced to a new line of mowers. The Yard Boss Lawn & Garden Equipment was the new line built by MTD that wore the silver and charcoal colors of the new line.

Jacobsen was building machines for Oliver at the same time it was building them for Ford and itself. The machines were basically identical, but they had obvious differences in the sheet metal as well as in the hoods and platforms. The larger Ford models had a different steering wheel and center cap. The characteristic that most easily identifies a machine as an Oliver is the original serial number tag.

The Trend Truck

White Motors had been in the truck business since the early 1920s. In 1957, it acquired the REO truck company of Lansing, Michigan. White also acquired the Diamond T truck company of Chicago, Illinois, in 1958. White moved production of the Diamond T to REO's Lansing, Michigan, plant in 1960 to create the Lansing division of White Motors as part of White's effort to diversify. The companies operated separately in the beginning, but White eventually decided to start mingling products. The first thing it attempted to do was to push one of its new trucks into the farm market.

The Lansing division had created a new truck to fit into the medium-duty truck market (19,501 to 26,000 GVW). It was the fastest-growing segment of the truck market and had increased 300 percent in the last five years. The Trend truck was all new, and the cab was built of Royalex, a plastic-like, state-of-the-art laminated material owned at the time by US Rubber. It could be hammered and dented and still retain its shape. If the dent did not pop back out, it could be fixed with a 500-degree heat gun. When the Trend was shown at the Growing O shows and other demonstrations, someone would stand in front of the truck with a sledgehammer and beat on it to prove its durability. Royalex is still available today and is most commonly used on canoes to make them lightweight and resistant to dents.

Production began on the Trend truck in June 1966. It was originally marketed as a Diamond T truck and an REO. It didn't take long for White to realize that its Oliver and Minneapolis-Moline dealers could also help sell this truck.

White Motors was trying its best to market the Trend truck through every available outlet. They were sold by White Motors, Oliver, and Minneapolis-Moline dealers. Several variations of literature were printed for the truck's different applications. *Author's collection*

The average farm in 1966 consisted of 350 acres. It was predicted that the average size of a farm in 1980 would be 600 acres. With the growth in acreage and better farming methods, a farmer needed more than a gravity wagon to haul in the crops. The Trend truck was perfect for that job.

The Trend proved to be very competitive with other mid-sized trucks in production. Oliver territory managers were encouraged to market the truck through their farm dealers. The dealers themselves often bought a truck and had a tilt-bed installed. This proved to be a good advertising tool for many dealers as they delivered new and used equipment to their customers.

The truck was designed to be very versatile. White advertised that the beds could be changed in minutes. It would undoubtedly not be as simple a procedure, however, as unhooking from a gooseneck trailer.

The Trend was originally marketed by five different companies: White Trucks, Diamond T, REO, Oliver, and Minneapolis-Moline. In an effort to create a unified product line with reduced cost, White decided to consolidate the business of Diamond T and REO in 1967 and create a new truck called the Diamond REO.

The official Trend logo turned out to be just that—a trend. Like most trends, they all come to an end in a short amount of time. *Author's collection*

The 1967 annual report shows sales of the Trend as moderate. Heavier optional components and other increased options were added in an effort to grow customer acceptance for both farm and city use. This series would be known as the heavy-duty (HD) series. White was still optimistic and hoped to meet its five-year sales goal.

The very early models had a 318 Chrysler engine but were short lived. By the time the trucks were marketed, the engine had switched to a 327 Chevrolet. The truck still lacked power, so an optional 350 engine was offered. The 350 became the standard engine in November 1968. An optional 366 was later available for more power.

One of the options that made the Trend so handy around the farm was the short wheelbase. Six different lengths were available that ranged from 101 to 179 inches. The cab-over-engine and set-back axle made it very easy to get around corners and through fences

Oliver did its best to market this truck to farmers and used this artwork on promotional material sent out to its customers. *Author's collection*

on the farm. The Trend, with an 18-foot body, had a turning radius of 25 feet, 4 inches versus 36 feet for conventional models. The short wheelbase also had its drawbacks, however. Gene Kuehn, one of the key players in the Growing O shows, said he had the task of driving one of the Trends from a show in Washington state back to Charles City, Iowa. The only thing he claims to remember about the Trend truck is that it nearly broke his back and he never wanted to get in another one.

The Country Wagon was the most unique model in the series. It was set up to allow the mounting of a camper. A 20- or 22-foot camper manufactured by Century Trailer of Webberville, Michigan, was specifically designed to be hauled by the Country Wagon. The camper had the most modern conveniences of the era. It was equipped with a separate shower and toilet, 30-gallon water storage, 6-gallon water heater, four-burner Magic Chef stove/oven, forced air furnace, and wood paneling. Air conditioning was part of the many extras in the deluxe package. There is quite a bit of literature out on the camper specials, but only 24 of them were actually built.

It is estimated that a total of 1,500 Trend trucks were built. Early production estimates were that 20 trucks would be built per day, and Oliver would receive one fifth of those built. Oliver had contracted three manufacturers to install tilt-bed and rollback beds on the dealers' trucks. The manufacturers were Schwartz in Lester Prairie, Minnesota; Mid-Equipment Corp. in Grundy, Iowa; and Henderson Mfg. in Manchester, Iowa.

Specifications for the HD Series:

HD 240: 24,000 GVW; 327 engine
(350 became the only engine available in November 1968)
HD 255: 25,500 GVW; 327 standard, 366 option
(350 became standard with optional 366 in November 1968)
HD 260: 26,000 GVW; 366 standard
HD 275A: 27,500 GVW; 366 standard (air brakes)
HD 275H: 27,500 GVW; 366 standard (hydraulic brakes)
White Trend Diesel: 27,500 GVW; Cummins 140 V-6 engine, 140 hp, 352-cubic-inch
White Country Wagon based on HD 255

The Royalex plastic cab would never rust, although the rest of the truck would rust. White thought this cab was the wave of the future, and it was still heavy even though it wasn't made of metal. *Author's collection*

The trucks seemed to be successful from a dealer's standpoint. The cabs were nearly indestructible, and since they had a Chevrolet engine, parts could be purchased anywhere or worked on in any shop. The 327 was short on power, a downfall that was remedied by the 350 engine. The optional Eaton two-speed axle was something they all needed in order to run faster than 50 miles per hour.

White decided to sell Diamond REO in August 1970, but it wasn't until mid-1971 before a buyer was found. Diamond REO was sold to F.L. Cappaert for $1.3 million in cash and $3.2 million in notes. This move ended all Trend production. The Trend was produced from 1966 to 1971, but most models found seem to be from the early production years.

The Trend was the only truck that White tried to market through Oliver dealers. The Oliver dealer had to have a special truck franchise contract and obviously didn't stock parts on the shelf when he knew only a few units would sell. Sam White Jr. summed it up best when he said, "We had no business trying to sell a truck!"

YOUR OLIVER DEALER OFFERS YO
$50
TRADE-IN VALUE
ON YOUR OLD MANURE SPREADER
ANY MAKE, ANY AGE, ANY CONDITION

CHAPTER 9

TELL 'EM AND SELL 'EM

"Tell 'em and sell 'em"—that was one of many phrases often used by Oliver when working with the sales department. The company wanted everyone to talk about its great line of farm machinery. As long as farmers were talking about Oliver, the company was getting free exposure. Word of mouth is always the best selling tool.

Oliver employed many different gimmicks through the years to sell its products. Some of the gimmicks were strange, but they got attention and that was the purpose.

Hart-Parr Tractor Schools

Marketing is a term that businesses today, unlike most companies at the turn of the twentieth century, are all familiar with. Still, many different techniques and gimmicks were used by early companies to raise awareness and make the potential user familiar with the product.

When gas tractor farming hit the scene in the early 1900s, farmers and mechanics were awestruck and didn't really know what to do with the new iron horse. Hart-Parr knew this was not good as operators took the expensive tractors to the field. Without proper operator training, the tractor was incapable of providing maximum performance and smooth operation. The tractors didn't look very desirable to own unless the operator knew what he or she was doing.

Hart-Parr decided that the best thing it could do was to train the dealers and mechanics on how to use the machines to their full capabilities; thus the Hart-Parr tractor schools were born. From 1909 to 1915, the schools taught owners how to use their tractors. The classes were often held at the branch houses or at the larger dealerships and proved to be very beneficial to the sale of Hart-Parr tractors.

Although the schools were successful, not everyone could travel the distance required to a branch house location. Hart-Parr started to offer a correspondence

In the 1900s and 1920s, dealers would take horses in on trade when trying to sell a tractor. However, the company preferred to take equipment in on trade.
Author's collection

The Oliver sales team for each of the branches was ever present at shows to help market and educate others about various Oliver products. From left to right: Ray Apple, Gene Kuehn, Ed Kruse, C.W. "Bally" Lloyd, and Bob Johnstone. *Author's collection*

The More Brothers became the Wimbleton, North Dakota, Hart-Parr distributors in 1903. A second location was opened in Fargo and sold hundreds of the large Hart-Parr models. The dealership conducted many tractor schools, such as this one in 1911. *Author's collection*

Hart-Parr's job wasn't just to build a good tractor. The company also felt an obligation to educate the dealers and distributors about the machines it sold. *Author's collection*

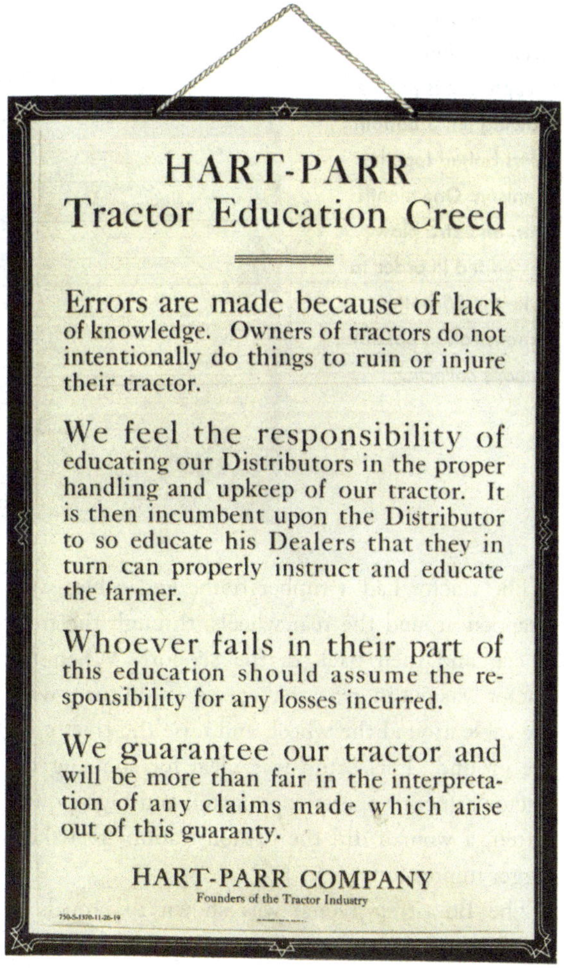

course beginning in 1911 for those who couldn't attend the class in person. The course consisted of fifteen different lessons ranging from drivetrain to fuel systems. The individual lessons were sent back to the plant and graded. This helped educate potential buyers and boost their confidence in making the big decision of buying a Hart-Parr tractor.

Oliver's Record-Breaking Plow Demonstration

While Hart-Parr taught men to drive and maintain tractors, Oliver showed off the ease of draft of its plows in a record-breaking demonstration. On October 14, 1911, the experts of the Oliver Chilled Plow Works teamed up with the experts of the Rumely Co. to put on a demonstration that promoted both the tractor company and plow manufacturer.

The Oliver and Rumely demonstration was held at Purdue University in Lafayette, Indiana, where seven carloads of machinery were brought in by rail from LaPorte and South Bend. The plow consisted of 10 five-bottom sections that were designed to link together and make one unit.

Oliver plow designer W.L. Paul and Rumely designer John Secor combined three 30-60 Rumely tractors and hooked them to fifty Oliver 14-inch bottoms to plow an acre in four minutes and fifteen seconds. Everything worked perfectly, and the demonstration, witnessed by students and professors, was flawless. The companies repeated the demonstration again on October 20 at the same location.

The purpose of the demonstration was to show the efficiency of modern farm equipment. The Rumely tractors ran on low-grade distillate, which cost four cents per gallon. The three engines consumed a total of 22 gallons an hour to reduce the fuel cost to six and a half cents per acre, and only a handful of men were needed to accomplish the task. If horses had been used, it would have taken fifty men and one hundred horses. They could have worked only ten hours before the teams would have had to be switched out. The tractors could have run round the clock with little down time other than refueling.

This demonstration caused such a buzz in the industry that International Harvester wanted to show its stuff too. IHC put on its own demonstration just outside of the Oliver plant in South Bend. It brought three 45-horsepower Moguls to the field and set out to beat the record of the Rumelys by hooking to a 55-bottom Oliver plow. This combination made a swath over 64 feet wide. The plow combination was similar to the earlier demonstration with the addition of one more 5-bottom section. According to newspaper articles of that time, "the Moguls carried the stupendous load easily and without a whimper."

Bootstrap Demonstration

Hart-Parr had its own exhibition of power with the bootstrap test. This display was first set up in the late 1910s to demonstrate the smooth operation of the new Hart-Parr tractor.

Oliver's photographer surely had fun trying to capture this picture showing ten 5-bottom plows bolted together in unison. One month later, an extra plow was added in order to make the 55-bottom demonstration possible. *Author's collection*

The tractor had a timber frame and cables were wrapped around the rear wheels, through the front wheel, and then back to the support. When the tractor was put in gear, the wheels would turn, wrap the cable around the wheel, and raise the tractor off the ground. This setup was ideal for showing the tractor's smooth operation, control, and high power. Often, a woman did the demonstration to make a bigger impression.

The Bootstrap Demo was shown to crowds at fairs and other events well into the 1920s. It was a popular and impressive exhibit of power that caught the attention of anyone walking by.

Oldest Tractor in Production

Hart-Parr wanted to show the durability and long life of the Hart-Parr tractor during the 1920s. The company sought out the oldest tractors still operating and bought them back from the original owners.

George Mitchell bought a Hart-Parr tractor in 1903 and used it as part of a threshing crew. It was used constantly until 1924 when Hart-Parr bought it from Mitchell for the scrap price of $72.75.

Hart-Parr prepared the tractor for display by removing the canopy and replacing it with a billboard. This signage boasted the machine was the oldest operating gas tractor in existence. This tractor became known as No. 3 because it was part of the third production model run. The tractor was shipped to many state fairs and displays around the Iowa area to show that it was the "granddaddy" of the tractor industry. Today, that same machine is still the oldest operating gas tractor in existence. It has been preserved, is fully operational, and is owned by the Smithsonian Institution.

The 70 Color Contest

Oliver kicked off the State Fair Exhibit Program in 1937. The company was experimenting with different colors and let the customers decide what they would be. Oliver put together six different color combinations and painted a tractor in each color scheme. The combinations consisted of:

Chrome green body, red trim, ivory lettering
Regatta red body, aluminum trim, white lettering
Chrome green body, tangerine trim, white lettering
Yellow body, black trim, red lettering
Chinese Gold body, tangerine trim, ivory lettering
Ivory body, Chinese Gold trim, red lettering

Each of the Oliver branches had a set of tractors in these colors. The branch closest to the state fair would take its tractors for the demonstration. Ballots were printed up, and passersby could vote on their favorite color combination. The ballots also included the voter's name and contact info and served as a prospect list. In exchange for the color preferences, the voters were mailed a nice leather key case embossed with the Oliver 70 logo.

This bootstrap test was demonstrated in 1919 using a Hart-Parr 15-30A. The fact that the test was done by a woman made the gimmick more impressive to those observing. *Author's collection*

With over one hundred thousand votes, the winning color was chrome green with red trim. Tractors that had been painted in that combination were sold by the branches for ten dollars over retail. One might assume they would have been sold for a cheaper price, but Oliver wanted to capitalize on the special paint scheme by selling the special edition tractors.

National 70 Days

The Hart-Parr 70 was introduced in October 1935. The new streamlined Oliver 70 with the "voters' choice" paint scheme was introduced in October 1937.

Oliver teased the public with the secretive introduction of the new 70 for two weeks in 1938. Oliver dealers all over the country whitewashed their windows on October 4, 1938, to hide the 70. A poster was put in the window that told everyone to mark their calendars for October 15 and 16. The following week another poster was placed in the window, announcing America's newest tractor would be shown soon. If dealers had a new 70 in stock, they were instructed to cover it with a white sheet or canvas cloth. On October 15, the windows were cleaned to unveil the new, streamlined 70. The dates ran one week later—October 29 and 30—for the Canadian dealers.

Oliver celebrated Oliver National 70 Days, which started on the launch date and lasted 70 days. Dealers were encouraged to sell as many 70s as they could before the peak spring season. Every window in the country was extensively decorated to draw attraction to the shiny new 70. The campaign worked, and the end result was a tractor with the largest production numbers of any other Oliver model.

Raydex Razor Blades

In 1939, Oliver introduced the new Raydex plow points, comparing their edge to that of a razor blade. The razor's edge comment was made enough that someone in the sales department thought it would be a novel idea to offer an actual Raydex razor blade to help promote its new replaceable shares.

Oliver teamed up with Gillette to package razor blades in a special Oliver wrapper. When the blades were ordered by the dealers, the wrapper would have the name of the dealer, but 250 packages had to be ordered at the bargain price of $6.25. This new item

At the 1924 Minnesota State Fair, Hart-Parr went all out to draw a crowd. In addition to the Bootstrap Test, it also had Hart-Parr No. 3 on display as the "Oldest Operating Tractor In The World." This tractor was shipped via rail to many different Hart-Parr displays before it was donated to the Cooperstown Museum in 1960. It is now the property of the Smithsonian Institution. *Author's collection*

was a great giveaway to customers or at fairs with a minimal investment.

The principle of the razor blade was the same as the plow point. You used them until they were dull and you threw them away. The wrapper on the razor blades was designed to match the box of the plow points. "Shave with a blade. Plow with a Raydex."

$50 Manure Spreader Trade-In

Oliver offered a special rebate when it introduced the new No. 7 rubber-tired manure spreader. The regular price of the No. 7 Superior spreader was $272, but Oliver was willing to take any spreader in on trade. It didn't matter what brand of spreader, how old it was, or what condition it was in—it was worth fifty dollars in trade-in value. Oliver was also willing to make payment arrangements using the Oliver Easy Payment Plan. This was a great introductory offer for the No. 7 spreader, but it only lasted from July through October 1939.

Diesel Economy Program

Oliver began a campaign known as the Diesel Economy Program in 1954. In order to sell the more expensive diesel tractors, Oliver offered to pay for half of the fuel consumed during a six-month time period. Users who purchased a new diesel tractor were given a record book to record the total number of hours the tractor was operated, total gallons of fuel used, and total cost of the fuel consumed in the six-month period. Oliver would issue a refund for the cost of half the fuel upon completion of this record book for the six-month period. The new tractor owner was to include the delivery certificate in the book as proof of ownership. This campaign helped Oliver sell many diesel tractors to those skeptical of the higher-priced tractor. The program expired on December 31, 1954.

The Oliverettes

The Oliver Farm Equipment Company always looked for a way to attract attention to its products. When Oliver introduced its new models for 1960, it wanted to add a little something extra to the show. The 1800 and 1900 checkerboards were the new models introduced late in 1959. Oliver was still trying to market some of the older three-digit models and wanted to make sure they still received some attention and weren't overshadowed by the new models.

The sales department wanted to put focus on the three-digit models' ease of handling and showmanship.

SAFETY RAZOR BLADES HELP SELL RAYDEX BASES AND POINTS

Grandpa used a straight edge razor—had to strop and hone it regularly. Grandson uses a safety razor, throws away the blade when it becomes dull and puts in a new inexpensive factory-sharpened blade.

The principle is the same with Raydex Plow Points. Use 'em until they're dull—then throw 'em away and put on brand new Points—which cost no more than what a blacksmith charges to sharpen an old-fashioned share.

There, in a nutshell, is the selling story on Raydex. And to demonstrate this comparison—Oliver offers advertising novelty packages of three fine double-edged razor blades—at remarkably low prices. The blade box is designed in miniature to resemble the Raydex Point economy package. *Your name and address is printed on one side of each package.*

A minimum order of 250 *imprinted* packages—750 razor blades—will cost you *only $6.25.* Why not send for a supply today—to give away on Oliver Days and Raydex plowing demonstrations. Here's a useful, cheap novelty which ties up perfectly with the product it advertises.

PRICES—RAYDEX RAZOR BLADES

250 pkgs.	(750 blades)	$ 6.25	1000 pkgs.	(3000 blades)	$22.75
500 pkgs.	(1500 blades)	11.75	2000 pkgs.	(6000 blades)	44.00

Order through your Branch. Be sure to indicate package imprint desired.

In order to get the message across that the Raydex edge was as sharp as a razor blade, Oliver teamed up with Gillette to sell specially packaged blades with the dealer's name on them that would be handed out to customers.
Author's collection

With this in mind, it came up with a very unique presentation. Oliver realized it was dealing with a male-dominant crowd and felt it could get attention if women were involved in the presentation. George Bird, the Charles City plant manager, came up with the unique idea of having flashily dressed women driving around on Oliver tractors in a dancing formation.

Plans for the tractor dancing routine began in the spring of 1959. Several people were required to pull the routine together. Personnel manager Ed Kroft enlisted his wife, Marion Kroft, to be in charge of wardrobe. She designed and made the women's costumes. Walt Gardner, part of the Charles City sales staff, was appointed the technician and tractor-training director. Agnes Gilmer, widow of Charles City's design engineer Louis Gilmer, was selected to orchestrate the drill team.

Tryouts were held to select ten women for the honor of being a tractor performer. It was not necessary for the women to be associated with the Charles City plant or Oliver; they only needed to know how to handle a tractor and smile at the same time. Two of the ten women selected would act as alternates since only eight tractors were part of the drill.

Training began with the basics: driving forward, backwards, stopping, starting, and shifting smoothly and quickly. Once they had these motions down, barrels were set up to act as an obstacle course. Walt and Agnes choreographed the tractor dance, and the women practiced tirelessly to refine the routine.

The tractors used by the Oliverettes were all Oliver 770 Industrials that were painted green. The Reverse-o-Torc transmission was needed to keep the routine smooth and simple, and the transmission was only found in the industrial models.

Once the routine was perfected, it was show time. The Oliverettes put on a 20-minute dazzling performance in front of crowds of Oliver dealers. Dressed in their Oliver green satin-fringed skirts, white gloves, and cowboy hats, the girls left the spectators in awe with a flawless display using beauty and user-friendly tractors. Around 3,500 spectators witnessed the performance of the Oliverettes during the course of twelve performances.

Put $300 in Your Pocket

By 1964, Oliver offered the 1600, 1800, and 1900 in the 100 series tractors. The company still sold the 550 and 770 tractors. In order to boost sales of the 770 models, Oliver initiated a campaign that offered $300 toward the purchase of a 770.

Vice president and general manager Don Koegle sent out a special brochure to Oliver customers and offered to put $300 in their pocket if they purchased a new 770. According to the letter, this offer was valid for only a short time. The listed retail price on a 770 in 1964 was in the $3,200 to $3,500 range. This

The Oliverettes demonstrated the easy handling of the Oliver tractors in front of crowds of Oliver dealers during 1959. This tractor ballet was only performed for one winter, but it was considered a huge success. *Author's collection*

bargain offer was nearly a 10 percent savings on the cost of a tractor, which was already a bargain-priced tractor compared with the newer models.

In addition to the $300 price reduction, Oliver also offered its Pay-As-You-Produce Plan, which allowed payments to be made as crops were harvested and eliminated the need for payments when there was no income.

Certified Power

The term "certified horsepower" was introduced along with the 50 series in 1965. Every tractor that left the plant had a sticker on the left side panel to designate the minimum PTO horsepower rating of the tractor.

Dealers argued that they couldn't make the customer understand that they were posting the minimum horsepower and not the rated horsepower. In response to this misconception, Oliver sent out brochures and had articles in nineteen different trade publications to educate the consumer on the difference.

Occasionally tractors were tuned up by other manufacturers when they were taken to the Nebraska testing grounds in order to get the highest rating out of it. Every adjustment possible was made to acquire a better horsepower rating to note in the record books. For example, tractor A was tested with all components set for peak performance during a short time span. The tractor ended up with a 78-horsepower rating.

In order to meet industry requirements, all tractors were put through rigorous testing at the University of Nebraska test lab. Tractors were hooked to this machine and tested for fuel consumption, slippage, horsepower, and many other ratings. *Author's collection*

When the dealer attempted to sell a production model of tractor A with factory settings, there was no way it could put out 78 horsepower, and the new owner was left with a tractor that could only produce about 70 horsepower.

Oliver's remedy to this problem was to guarantee the tractor would produce the horsepower specified on the tag. The company made certain changes in the engineering and fuel system to establish more control over the horsepower output. Each tractor was put on a dynamometer as it rolled off the production line. That rating, along with the initials of the inspector, went on the decal. If the tractor could not develop the minimum horsepower rating of that model, it didn't go to the dealer.

Horsepower ratings increased as the tractors were used due to less internal friction. The certified horsepower rating gave the buyer confidence because he knew it had been tested accordingly and had the actual rated horsepower. This certified horsepower rating became the standard no matter what other tests claimed and did a lot to boost the confidence of both the buyer and seller. The program was so successful that the engineering and production department received a special award from President Sam White, Jr.

Free Plow

The Oliver 2655 was the same tractor as the Minneapolis-Moline A4T-1600. Sales were not strong for this model, and the sales department knew it needed help to reduce the inventory on hand. The last year that the tractor was produced in Minneapolis was 1972.

Compared with other brands, the 2655 was a little high priced, so White Farm Equipment reduced the price $2,000 in May 1972 in order to make the tractor more competitive. It was a good sales boost, but the company wanted more.

Every person who purchased an A4T-1600 or 2655 in September and October 1972 received a free model 548 six-bottom cushion trip plow or the equivalent list price value on a companion implement of his choice. This tractor sold for about $16,000 before the $2,000 reduction, and the plow discount was $2,000 as well, which resulted in a 25 percent discount off the price tag.

Certified
OLIVER CORPORATION CERTIFIES THAT THIS TRACTOR SERIAL NO.__________WILL PRODUCE AT APPROXIMATELY 1000 FEET ALTITUDE ABOVE SEA LEVEL WHEN NORMAL BAROMETER AND 70° TEMPERATURE PREVAILS AND WHEN ENGINE IS TUNED CORRECTLY AND BROKEN IN ADEQUATELY, AT LEAST_____OBSERVED P.T.O. HORSEPOWER.
____
INSPECTOR

GM wasn't the first company to certify its vehicles. Oliver certified each tractor individually during this era, and each sticker was initialed by the employee who performed the test. *Author's collection*

Growing "O" Shows

In 1963, the traveling road show, known as the Growing "O" Show, hit the circuit to introduce all the new Oliver products available. In a 33-day span, the tour hit nine of Oliver's branch cities where they put on the show of shows. Dealers and their customers were all invited to this well-orchestrated show where Oliver equipment was not only displayed, but also used in actual field conditions.

Semi trailers that hauled equipment in to the shows were converted to mobile bleachers. Instead of walking from field to field and watching the equipment go by, the crowd was hauled alongside the equipment as it operated to get a very close look at its performance. A lift was mounted on the back of a tractor that included a speaker system. One of the Oliver reps would speak about the products alongside the crowd as things were in action.

Oliver believed that demonstration was key to sales, and the Growing "O" Shows proved it. After the first year, sales increased 7 percent. The next year showed another 17 percent increase and at one point, sales increased 28 percent. It cost Oliver around $80,000 in the 1960s to put on each show, but the increase in sales seemed worthwhile. Although Oliver continued to demonstrate at Farm Progress Shows and other events, the stand-alone road show put on by Oliver came to an end after the 1967 season.

OLIVER 66
OLIVER 77
OLIVER 88
OLIVER T
OLIVER R
OLIVER A
OLIVER C
OLIVER T
OLIVER O
OLIVER R
88 TRACTOR
NEW
OLIVER
77

CHAPTER 10

THE COMPANY

In the early years of the agriculture industry, every one of the companies under the Oliver flag was a giant in its field. The Oliver Chilled Plow Works was the Plowmaker to the World. Hart-Parr was known as the Founder of the Tractor Industry. Companies like American Seeding, Nichols & Shepard, and Farquhar were successful and well known in their fields of expertise. The union of all these manufacturers should have put the company at the top, but Oliver never could attain the highest market share.

C.R. Messinger came to Oliver as chairman of the board in 1930. The company was $17 million in debt to banks at the time. Messinger was not an implement man but was brought in by the banks to get the company's finances back in order.

The company had a great product line that put some of its strongest competition back on its heels. But the competition came at a price, however, because the company had to gamble on future credits.

The company would do anything to sell a tractor; it would even take horses in on trade. Before long, Oliver had thousands of hungry horses that needed to be fed and ate away any potential profit. The horses had no value to Oliver and became liabilities rather than assets.

Messinger demanded the company tighten its belt however it could in an effort to reduce debt. He talked to every department head and told them to create an economical budget. Of course, each department wanted the cuts to come out of other departments. Messinger looked at all the proposed budgets and threw them all in the trash. His solution was a reduction in dollar percentage for each department. It was up to each department to decide how the cuts would be made. The departments had to decide internally whether they would reduce personnel, reduce salaries, or cut expenditures some other way.

The plan to put Oliver back in the black worked, and in 1936, the company showed its first profit

In 1948 when Oliver introduced its new Fleetline series, the fair tents were a popular attraction. Oliver went all out with banners, literature, and even a rotating tractor with spotlights on it such as this one at the Iowa State Fair. *Author's collection*

since its formation in 1929. Net profits for 1936 were $1,153,636, compared with a loss of $482,005 the previous year. Sales had increased 53 percent, and much of the credit for the increase in sales was given to the Hart-Parr Oliver 70 introduced in 1935.

The Oliver Farm Equipment Company reached a milestone on October 31, 1938, when it paid off all its bank loans. The $17 million loan along with $4.5 million in interest was wiped from the slate. With the payment of the last note, Oliver held a mortgage burning party at the Hotel La Salle in Chicago. Branch houses held parties at their locations. A speech given by C.R. Messinger was broadcast live to each of the branch parties via direct telephone hookup. It was an occasion celebrated by all who played a role in freeing the company from debt.

Oliver was ready to run. It had a modern, stylish line of equipment, and the future looked bright. Upon the announcement of the mortgage burning, Oliver rehired 350 Charles City employees that had been let go to bring the employee number up to 900 at the Iowa plant.

On February 4, 1941, C.R. Messinger died suddenly from a heart attack at the age of 57. The man who had financially brought Oliver back to life was gone, and the United States was on the brink of entering World War II.

The Office of Production Management issued an order in late 1941 to limit farm equipment production to 80 percent of the previous year's production until October 1942. Materials were needed for war production, and many plants were converted to manufacture products to assist in the war efforts. Even with the reduction in farm equipment production, Oliver still achieved net sales of over $28 million, the largest in the company's history at that time.

In 1943, the War Production Board limited production of farm equipment to 20 percent of the amount

During the Fleetline era, business was good and dealerships took on a whole new look. The little shop out back was gone and in was the well-polished showroom and parts department. *Author's collection*

This picture was taken in late 1940 after the introduction of the Fleetline series. Oliver encouraged its dealers to clean things up and make a better impression. Even the man behind the parts counter is wearing a bowtie. *Author's collection*

produced in 1940. This was a severe cut in the production of ag products and in profits for Oliver. War products composed 62 percent of the sales for the year and kept the company operating at a profit.

The production limitation order was lifted in 1944, and the company could again produce farm equipment up to 80 percent of the 1940 production level. Oliver acquired the manufacturing and production rights of the Ann Arbor Machine Company in an effort to expand its farm equipment line.

The Oliver Farm Equipment Company acquired the crawler business of the Cleveland Tractor Company at the end of its 1944 fiscal year. Net sales at the end of 1944 were once again at a record high with over $43 million in sales. Special war products consisted of 44 percent of that amount and were subject to renegotiation.

Profits decreased in 1945 due to low sales, but the company still forged forward with a costly plant modernization program and an ambitious program for developing new products. Strikes at several of Oliver plants during the following year caused the company to operate at a loss for the first half of the year. Strikes at the plants of many Oliver suppliers also caused setbacks. Without the materials to build equipment, the company was crippled.

The engineering department was hard at work and designed a new fleet of tractors. When they were introduced in 1948, the success was reflected in a 103 percent sales increase over 1946 sales. This superbly designed tractor sent the other companies running to the drawing board in an effort to keep up. Oliver was proud to say that it had been operating at a profit for thirteen consecutive years. Confidence was running high, and the company borrowed $8 million for working capital.

Several programs of major capital expenditures continued in 1948. A new branch house was built in Memphis, Tennessee. The South Bend plant built a new shop, and Charles City constructed a new building to house a new moving-belt floor-level assembly line for the new Fleetline series of tractors. Oliver had 9,100 employees in the plants, sales branches, and general offices at this point in time.

Strikes throughout the industry caused many production interruptions in 1950. The Cleveland plant was shut down for six months. Oliver initiated wage increases, pension plans, and retirement plans in an effort to stabilize employment and increase employee efficiency.

Oliver also created the Oliver Suggestion Plan, which was designed to incorporate employee ideas into savings for the company. Employees were urged to turn in ideas that might increase efficiency on the production line or improve the company's products or operations. Employees were monetarily compensated for their ideas according to their usefulness. A simple suggestion could be worth from $5 up to $4,000. It was a chance for the employee to feel more like one of the company. If someone was able to make improvements and receive full credit for it, he or she was always thinking of ways to contribute. A typical suggestion might be changing the grey iron drag chain ring to one made of steel or moving two boring machines closer together to combine operations of the selector valve body used on the Hydra-Lectric unit.

Oliver went into an aggressive advertising campaign with the Fleetline series in production in order to make up for the losses caused by earlier strikes. Two of its advertising attempts in 1950 received awards from the National Advertising Association. The Best National Advertising of the Year award went to the ad focused on Oliver's new Bostrom seat. The second award was part of the 100 Best in Outdoor Advertising and featured a billboard display of an Oliver 77.

When Hart-Parr and Oliver teamed up to design the new series of tractors introduced in 1930, it turned to Waukesha to supply the necessary power using a VIS engine. Oliver eventually purchased stock in the Waukesha company to help add to its profit margin. *Jeff Hackett*

Of the Fleetline series, the 66, 77, and 88 were built as orchard models. Styling became an important part of tractor designs after World War II. This series of tractors was known as the "three beauties" and was undoubtedly one of the most popular styles of tractor in that era. *Super T*

Oliver had ten thousand employees by the end of 1951. It had purchased another manufacturing plant in Battle Creek for building Boeing fuselages. It was also on the heels of acquiring the A.B. Farquhar Company of York, Pennsylvania, which had valuable government contracts.

The sale of defense products had increased to over $46 million in 1953. It was a sharp increase from the $17 million sold in 1952. Even though the defense products were bringing a good income to the company, the sale of regular products had dropped from $117 million to $89.9 million. The decrease was blamed on the poor prices of farm products and a widespread drought. Oliver expected a turnaround in the near future and worked to develop new, improved products to stimulate sales. The result was the Super series of tractors.

Oliver had been looking for a manufacturing facility on the West Coast for several years. After a two-year negotiation, Oliver finally purchased all of the outstanding capital stock of the Be-Ge Manufacturing Company, which was in the midst of filling defense contracts. Be-Ge operated as an independent division of Oliver and retained its previously established dealers. It now had additional dealers, however, thanks to Oliver's distribution system.

The drought that threw a dark shadow over the farm industry continued in 1954 and caused a drop in net sales. This is where many believe Oliver started to slide in the wrong direction.

Sales for 1955 were up slightly but still lower than two years prior. Oliver was forced to borrow more money in order to increase the winter inventory in preparation for the busy spring season. In addition, Oliver's aviation division had lost its government contract, and the potential of finding a replacement of that caliber was slim.

Oliver had spent over $48 million in the last ten years for property additions. Its net worth was over $77 million, but the company anticipated a further decline in income due to the poor farm economy. Oliver knew that the economy wasn't something it could change and looked at its other divisions to see how it could make up for the loss in the ag division. Oliver entered the outboard motor market at this time. The second plant in Battle Creek was sitting empty, and the government was giving out incentives to industries that would give a boost to recreational activities. It seemed like a win-win situation, but in all reality, it was a lose-lose deal. At this time, Oliver also owned 11 percent of the Waukesha Motors Company, with whom it had a close working relationship.

Oliver continued to slide downhill in 1956. Customers sharply reduced their farm equipment purchases due to the uncertainty of federal farm programs to aid drought-stricken farmers. This lower sales volume increased manufacturing costs since materials were no longer ordered in such large quantities. Sales were not good in the United States, but export sales were up. Oliver looked for another market and established Oliver International, S.A., to handle the foreign sales.

Oliver was hoping for a better year in 1956, so it built up its inventory during the winter of 1955. Products were built, and raw materials were ordered for future production schedules. It soon became apparent, however, that 1956 was not going to be a better year and something needed to be done. The company's money was tied up in products sitting on lots with little chance of a quick turnaround. Oliver began an expense and inventory reduction program. Employment was decreased from 9,383 employees in 1955 to 5,557 one year later. Sales were at a six-year

During the late 1940s and early 1950s, Oliver built several new branch houses to help with distribution. This was the new branch in Kansas City, Kansas. *Author's collection*

The 77 and 88 were produced in the highest numbers of the Fleetline series. These tractors were designed to share many of the same parts, which helped reduce needed inventory for the dealer. *Super T*

low, and net earnings per common share dropped from $2.13 to $0.76.

The ship was starting to sink, and the board was looking for a lifeline. J.I. Case was also experiencing financial difficulties. It was looking for a merger with another successful company in order to satisfy its creditors. In August 1956, Oliver's chairman of the board, Alva Phelps, announced that discussions were being held between Oliver and Case regarding a possible merger. Within a month, Phelps announced that the talks had been terminated. Case eventually merged with the American Tractor Corporation.

The great steel strike of 1956 was over but had resulted in a substantial increase in the price of steel. Oliver was already operating at a loss and had no margin left to cushion this blow. Prices on the next production of tractors were sure to increase. Dealers were encouraged to buy all they could at the time in order to have the lowest-priced tractor on the lot.

Oliver had been involved in a diversification program for several years. What was once a farm equipment company was now involved in the outboard motor industry, manufacturing in foreign countries, and defense products. Unfortunately, not all of these aspects were moneymakers. Total earnings for the 1957 fiscal year were one third of those in 1956. Losses suffered in the other markets reduced earnings from the sale of farm equipment. Sales of the outboard line had nearly doubled, but it operated at a loss due to the high price of development costs.

The net earnings of a common share were now down to $0.13. The earnings per share were drastically smaller than the $4.69 per share in 1948. Stocks were split once during that time, which made the picture a little brighter than it appeared.

The one constant with Oliver was the profit from defense products, but even that changed in 1958. Oliver's defense operations were operating at a loss, and the outboard line had created a constant drain for the company. Oliver decided it was time for an extensive companywide program of cost reduction and profit improvements.

The first thing on the chopping block was South Bend's No. 2 plant. The manufacture of the products in that plant were moved to Charles City, the most notable being the Super 44 and the Super 99. Oliver sold off its share in Waukesha Motors and brought an additional $1 million in working capital back to the company. Two new branch houses were built to replace the older, outdated, inefficient buildings.

Sales were expected to rise in 1958 due to an aggressive demonstration campaign put together by the sales department. The Springfield, Ohio, plant was shut down in 1959, and production of those products was divided between South Bend No. 1 plant and the Shelbyville plant. The outboard motor line was sold to Perkins for an exchange of stock. Oliver dealers could still sell the line, but the company didn't have to eat the production cost.

Sales numbers were starting to head upward, but it was too late. Corporate raiders had already slipped under the crack in the door. Two brothers, who were known to specialize in finding companies with good products and assets that were experiencing problems, had purchased enough Oliver stock to gain two seats on the board. They were soon added to the finance committee, which was like giving them the key to the safe.

Another element to the future destruction of the company came when a Chicago lawyer was added to

This was the scene of the major assembly line for final completion of the main fuselage section of the RB-47E Reconnaissance Airplane. The intricate assembly of this huge section, with its thousands of precision parts, gained recognition for Oliver's Aviation Division as one of the nation's most capable producers of defense products. *Author's collection*

the board. A new policy was adapted that automatically eliminated directors on the board after their seventieth birthday. That policy immediately eliminated two long-standing directors—C. Frederick Cunningham, the son-in-law of J.D. Oliver and one of the last Oliver family members with the company, was one of those casualties. This policy did not sit well with the current board and caused two others to submit their resignations. The two brothers were paving the path to a sell off of the asset for a nice profit and eliminate anything or anyone that might stand in the way.

The brothers insisted that they were merely investors but it didn't take long for them to start talking merger. The first was with Studebaker, who had been a longtime friend of the Oliver family. Negotiations only lasted a couple months before they were cancelled, but it didn't take long to find another interested party.

The White Motor Corporation was much like the brothers. It liked to trade off assets for a profit rather than build a good product, and it was in a financial position to do that thanks to the recent trucking boom.

Oliver became a subsidiary of the White Motor Company on November 1, 1960. At the time of the takeover, Oliver employed 5,800 men and women and had 2,100 dealers in the United States. Oliver International had 110 distributors in 93 countries. Industrial products were sold by 80 heavy-line distributors and 280 light-industrial distributors.

White only bought select assets of the company. According to the agreement, Oliver shareholders agreed to sell the physical assets of the farm equipment business to White. In exchange, they would receive 655,000 shares of White stock valued at $48.50 per share along with a cash payment. White also took the Oliver International export division, but it did not take Oliver Australasia.

White's acquisition did not include the Cleveland crawler plant, the crawler line, Be-Ge, or the Farquhar plant. White agreed to pay $200,000 a year rent for the Cleveland plant. After two years, White could buy the crawler plant's inventory at 100-percent book value and machinery/equipment/tooling for 80-percent value. White would become the sole distributor of the crawler line in the United States, but if White didn't have $12 million in sales the first year, Oliver could cancel its option to sell the crawler line to White.

The three-point hitch was first introduced on the Fleetline series. The combination of Hydra-Lectric, Bostrom seat, and three-point made this a fully loaded machine capable of doing any task on the farm. It was the modern options like these that made Oliver such a popular tractor. *Jeff Hackett*

Under the terms of the agreement, White was entitled to the Oliver name after the closing date, and the leftovers of the Oliver Corporation were required to change its corporate name. The Oliver Corporation became the Cletrac Corporation.

The Cletrac Corporation, now owned and operated by the two brothers, wanted to get rid of excess baggage. The York plant was closed upon the completion of the post office conveyor contracts. Select products were sold off, the plant was torn down, and the land was given to the city.

In 1961, the new Oliver Corporation purchased part of the crawler line at the earlier agreed upon sum of $6.5 million. The company wanted to keep the rights to produce the OC-4 and OC-9 series even though some exceptional new machines were ready for production. The remainder stayed with the Cletrac Corporation, but about the only product left was Be-Ge.

The brothers were buddies with some of the directors of the Hess Oil Company of New Jersey. Hess really wasn't interested in anything that was left of the Cletrac Corporation, but it did want a seat on the New York Stock Exchange, and the purchase of Cletrac would provide that. What was left of the Cletrac Corporation was absorbed by Hess Oil. Hess turned around and spun Be-Ge off to Victor Fluid Power.

Meanwhile, White looked to add more to its portfolio. In 1962, it purchased the farm implement business from Cockshutt Farm Equipment Ltd. for $8 million in cash and notes. It was an underhanded deal that still stings those affiliated with the Cockshutt company. White discontinued production of the Cockshutt tractors, took over its combine plant,

The 1958 line of tractors and equipment were introduced to the Portland, Oregon, branch family of Oliver. The well-rounded display covered the tractor line from the Super 44 to the mighty Oliver 990 and even included outboard motors. *Author's collection*

This early 1960 picture is typical of the small-town dealership of that era. When White Motors entered the picture, it pushed big new steel-structured buildings to its dealers. Some of the dealers held on, but many dealerships changed hands at this time. *Author's collection*

and moved Oliver's harvester division to the Brantford, Ontario, factory. The old harvester plant in Battle Creek was torn down, and the property was disposed of.

White Motors added Motec Industries (Minneapolis-Moline) to its stable. The book value for the company was $31.9 million, but White purchased the company for $19 million cash. MM was to operate independently with its own management and dealer organization. White's goal was to broaden the company's participation in the farm equipment market.

With three farm equipment companies in the family, White set out to consolidate and streamline operations. Minneapolis-Moline moved all its manufacturing to the Minneapolis plant and used the Hopkins, Minnesota, facility as a warehouse and an engineering and general office location. The Brantford plant was added to meet the demands of building combines for three companies.

White Motors attempted to consolidate Cummins and White to form a new company in 1963. This consolidation could have been interesting if it had taken place. Imagine an Oliver tractor equipped with a Cummins engine coming off the production line! But this was not to be. The Department of Justice opposed the consolidation and believed it would be a problem with other truck companies that used Cummins engines.

Since White couldn't buy an engine company, it decided to develop its own. White established the Advanced Product Division in Torrance, California. An entire staff was hired to develop and test a new line of engines that could be used in both tractors and trucks.

Meanwhile, in Charles City, the production of Oliver's successful crawler line was phased out. White had denied the necessary funds for research and development and put the crawler behind in the market share. The year 1965 marked the end of an era for the crawler models that had been in production for nearly fifty years.

White's Farm Equipment division consisted of Oliver, Minneapolis-Moline, and Cockshutt and had sales of $187 million, which was 24 percent of the total sales of White Motors. Sales seemed to be good for all the divisions, but White couldn't stop

Oliver had its own rail line out of all of its plants to help expedite equipment delivery to branches. Some dealers, such as this one, had its own side rail to unload equipment right onto the lot. *Author's collection*

spending money. In 1966, it bought 75 percent of the Arbos combine company in Italy. It purchased the assets of the Hercules Engine plant, and it also built a new plant in British Columbia for the production of White's Western Star. The next year the company acquired the remaining 25 percent of Arbos, which made White a 100-percent owner of the Italian harvester company.

White's sales reached a record of $851 million in 1968, yet the net income had a decline of 20 percent. White blamed the drop in earnings on the farm equipment division's low sales. White still continued to spend money. It purchased the Euclid line of equipment from GM and began construction on a modern, state-of-the-art engine plant in Canton, Ohio.

Sales of the farm equipment division were disappointing in 1968, and the result was an overstock of dealer inventory for the first time in six years. The following year wasn't much better. The farm industry was going through a depression. The demand for farm equipment was reduced by poor crop prices, rising costs, and high interest rates. Sales from White's farm equipment division were reduced 26 percent in 1969.

The farm industry was going through rapid changes. Farm ground was being gobbled up by growing suburbs, and the number of farms was decreasing. Those who continued in the business were farming more acreage and needed larger, more efficient machinery.

White decided it needed to take aggressive steps to restructure its farm division to change along with the industry. The result was the consolidation of Oliver

The Fleetline years of Oliver were some of the most successful. Oliver initiated a massive advertising campaign that used radio, billboards, and clips at movie theaters to introduce the new line of Fleetline tractors. *Author's collection*

and Minneapolis-Moline to form a new subsidiary known as the White Farm Equipment Company. White Farm's offices were consolidated to Hopkins, Minnesota. The dealer network was overhauled and had fewer dealers with larger territories and financing programs to meet the needs of the farmer.

This consolidation resulted in a fierce battle between the top management of Oliver and Minneapolis-Moline. Offices had been established at the home of MM. Oliver was seen as invading its turf, and the resulting conflict was not pretty. When the dust settled after the fight, the MM executives had won. It had nothing to do with who could run the company in the best manner but was all about politics and power.

Analysts didn't expect the farm economy to be much better in 1970, so White forged ahead with its reduction of dealers and their inventories. This didn't stop the company from spending money, however, and White acquired Alco Engine and Alco Products Service, Inc.

Half of the Charles City plant employees were laid off due to the continued farm depression. It was also announced that Oliver's most successful little plant in Shelbyville would phase out production, and the plant was shut down.

White's 1970 annual report stated that it was a difficult year marked by a weak economy, depressed

These signs began to appear in 1974 during the transition from Oliver to White. Although White was up to something, it wasn't enough to keep the Farm Equipment Division of White Motors alive and well. *Author's collection*

farm market, lower truck sales, and internal problems. White also stated that the farm group had been a major source of problems for the past two years. The farm division was accountable for only 10 percent of total sales by this time. Oliver's and Minneapolis-Moline's product lines were standardized, and an exchange of models between the two was started in an effort to reduce costs.

The next few years were like a game of musical chairs between Oliver and Minneapolis-Moline management. The seas were rough when MM took over management and attempted to do away with its sister company. When Oliver regained control, it attempted to do the same thing to MM. In 1972, the company was once again consolidated into one division known as the Oliver Farm Equipment Company, and its headquarters was moved back to Charles City.

All of these changes caused unrest amongst loyal Oliver customers. The company was no longer the stable, independently owned company they once knew. Rumors were circulating that White was ready to dump the farm equipment division. Many farmers started to look to other colors because they were not sure if Oliver would be around much longer. The Minneapolis-Moline plant was shut down, and production was transferred to Charles City.

The 2255 was the last tractor to roll out of the Charles City, Iowa, plant wearing the Oliver name and familiar Meadow Green/Clover White colors. From 1972 to 1976, 2,160 of these models were built. List price for the 2255 4WD in 1974 was $21,860. *Jeff Hackett*

In 1996, the American Society of Mechanical Engineers recognized the Hart-Parr tractor as a national historic mechanical engineering landmark. This plaque was presented to the city of Charles City, Iowa, the birthplace of the tractor. *Jeff Hackett*

The farm division experienced significant growth in sales for 1973. It was the first time the division had shown yearly earnings since 1968. Farm tractor sales had increased by 27 percent. Ironically, this was the same year that White started to phase out the familiar Meadow Green line of tractors.

According to White Motors 1973 Annual Report, "Farm Equipment traditions for tractor design and engineering were shattered through introduction of the White Field Boss tractor. In addition to a highly styled exterior treatment which reflects visually the pride which most farmers take in their equipment, the unit is also the first of its kind to be designed both for wheatland use and for crop farming." This was the first tractor to be marketed by the company in the United States under the White nameplate.

The battle among three different tractor companies would soon come to the end. It was obvious there would be no harmony if green, yellow, or red were used as the primary color for the new fleet of White tractors; therefore, a new color had to be chosen. The sales department's suggestion that the color reflect class and royalty influenced the choice of silver.

White had already eliminated the Cockshutt name, and the tractors and equipment sold in Canada had been tagged White for several years. By 1975, White was building an entire line of silver tractors. On February 13, 1976, the last 2255 wearing the Meadow Green color rolled off the assembly line.

The question "What happened to Oliver?" can't be answered in a few sentences. It can barely be answered in a chapter. It wasn't one thing that brought a familiar face to its knees but a series of events that caused a domino effect. The legacy of a farm equipment company that shared its name and products for over 120 years still remains today.

INDEX

Made in the USA
Monee, IL
01 March 2026

44705357R00090